00-WSM-7UR

STORMWATER THERMAL ENRICHMENT IN URBAN WATERSHEDS

by:

Mark S. Kieser
Jeffrey A. Spoelstra
"Andrew" Feng Fang
Kieser and Associates

William James
Yi Li
University of Guelph, Canada

2004

WERF

The Water Environment Research Foundation, a not-for-profit organization, funds and manages water quality research for its subscribers through a diverse public-private partnership between municipal utilities, corporations, academia, industry, and the federal government. WERF subscribers include municipal and regional water and wastewater utilities, industrial corporations, environmental engineering firms, and others who share a commitment to cost-effective water quality solutions. WERF is dedicated to advancing science and technology addressing water quality issues as they impact water resources, the atmosphere, the lands, and quality of life.

For more information, contact:
Water Environment Research Foundation
635 Slaters Lane, Suite 300
Alexandria, VA 22314-1177
Tel: (703) 684-2470
Fax: (703) 299-0742
www.werf.org
werf@werf.org

This report was co-published by the following organizations. For non-subscriber sales information, contact:

IWA Publishing
Alliance House, 12 Caxton Street
London SW1H 0QS, United Kingdom
Tel: +44 (0) 20 7654 5500
Fax: +44 (0) 20 7654 5555
www.iwapublishing.com
publications@iwap.co.uk

Library of Congress Catalog Card Number: 2004112636

WERF ISBN: 1-893664-82-1 IWAP ISBN: 1-84339-683-1

Kieser & Associates

ACKNOWLEDGMENTS

The project team would like to thank Mr. W. Christopher Barnes, P.E., City of Portage, for his support and cooperation throughout the project. Special thanks go to Ms. Nicole Ott, Mr. Brian Boyer, and Ms. Patty Hoch-Melluish of Kieser & Associates for their work in completing Appendix A of this report and for assistance in all aspects of this research. We are grateful to the Kalamazoo Foundation for providing funding that supported the initial purchase and installation of temperature monitoring equipment. We particularly would like to thank Mr. Dean Carpenter, this study's initial project manager at WERF and the five reviewers of this report for their valuable suggestions and comments.

Report Preparation

Principal Investigator:
Mark S. Kieser
Kieser & Associates

Project Team:
Jeffrey A. Spoelstra
"Andrew" Feng Fang, Ph.D.
Michael Crimmins, Ph.D.
Kieser & Associates

William James, Ph.D., D.Sc.
Yi Li
University of Guelph, Canada

Project Subcommittee

Water Environment Research Foundation Staff

Director of Research:	Daniel M. Woltering, Ph.D.
Senior Project Manager:	Margaret Stewart

ABSTRACT AND BENEFITS

Abstract:

Thermal enrichment of coldwater streams by heated stormwater during summer months is often overlooked. It can be exacerbated by traditional management practices that typically account only for flow moderation and pollutant removal. This study evaluated and identified innovative and traditional approaches to moderate this temperature impact by monitoring and analyzing the hydrologic and thermal regimes of an urban stormwater treatment system consisting of two traditional wet detention ponds and an enhanced natural wetland.

Data analysis showed temperature increases in the open detention ponds and the wetland's ability to mitigate this thermal enrichment. Event-based thermal loading and temperature regime analysis indicated flow reduction via infiltration and effective vegetative cover in the wetland were the primary mechanisms for mitigating stormwater thermal enrichment. Using the concept of temperature equivalent, the locations and potency of thermal enrichment source areas were established.

A heat transfer model was developed to simulate runoff temperature. Results indicated that rainfall characteristics, temperature difference between rainfall and the ground surface, and the runoff flow depth were the most important factors affecting runoff temperature.

Recommendations identify the need for additional mechanisms (flow reduction/ infiltration) beyond traditional detention ponds to mitigate thermal enrichment. Vegetative shading reduces direct heating of open channels. High albedo, or highly reflective, surfaces and shading reduce heat buildup in pavement, diminishing heat transfer to runoff during rain events. Underground structures that have contact with stormwater provide some limited benefits for reducing the temperature of heated runoff.

Benefits:

- Quantifies stormwater thermal enrichment in a typical urban setting
- Demonstrates the effect of stormwater best management practices on the thermal regime of runoff and the receiving stream
- Identifies contributing and mitigating sources of stormwater thermal enrichment
- Develops a thermal budget model mechanistically describing and simulating urban stormwater thermal enrichment
- Discusses management and design options necessary to mitigate urban stormwater thermal enrichment

Keywords: Temperature, urban stormwater runoff, coldwater streams, thermal enrichment, heat load, heat transfer, stormwater treatment, stormwater BMPs, detention ponds, wetland, infiltration, vegetative shading.

TABLE OF CONTENTS

LIST OF TABLES

LIST OF FIGURES

LIST OF ACRONYMS

ΔT	change in temperature
BASINS	Better Assessment Science Integrating Point and Nonpoint Sources
BMP	best management practice
CI	confidence interval
CrUSTe	Critical Urban Stream Temperature Model
EMC	event mean concentration
ha	hectare
HDPE	high-density polyethylene
HEATRAN	HEAT TRANsfer model
HSPF	Hydrologic Simulation Program – Fortran
K&A	Kieser & Associates
LULC	land use/land cover
m	meter
PCSWMM	PC Stormwater Management Model
QA	quality assurance
QC	quality control
RBPs	rapid bioassessment protocols
SD	standard deviation
SOPs	standard operating procedures
T	temperature
TE	temperature equivalent
TMDL	total maximum daily load
TSS	total suspended solids
TURM	Thermal Urban Runoff Model
U.S. EPA	U.S. Environmental Protection Agency
USGS	U.S. Geological Survey
WERF	Water Environment Research Foundation

EXECUTIVE SUMMARY

Millions of dollars will be spent over the next several decades implementing Phase II stormwater regulations, total maximum daily loads, voluntary watershed management plans, and habitat improvement programs. Mounting evidence suggests that stormwater mitigation and treatment approaches targeting only pollutant loading and flow moderation in urbanized areas are not sufficient to achieve overall desired water quality protection for coldwater habitats. Temperature, as one of the key environmental variables controlling aquatic biota, must also be considered in stormwater management planning where urban runoff is introduced into aquatic systems.

Thermal impact on coldwater streams by urban stormwater runoff is often overlooked, and enrichment can be exacerbated by traditional best management practices (BMPs). Coldwater streams in relatively undisturbed catchments typically respond to storm and runoff events with a decrease in temperature due to cooling short-term climatic conditions and an increase in groundwater contributions. Urban stormwater conveyance systems rapidly deliver surface runoff to receiving waters. Because coldwater streams are normally driven by groundwater inputs, this rapid delivery of surface runoff at any temperature above that of the receiving stream can be defined as stormwater thermal enrichment. Thermal enrichment of stormwater runoff occurs when: 1) stored energy is transferred from impervious surfaces to runoff; and/or 2) temporarily detained stormwater gains energy in conveyance channels and impoundments (e.g., wet detention ponds).

This five-year thermal enrichment study, supported by the Water Environment Research Foundation (WERF) and the Kalamazoo Community Foundation-Environment Now Fund, was completed for a recently constructed regional stormwater treatment system in Portage, Michigan. This Regional Stormwater Quality and Trailways Facility receives urban runoff from 187 ha of highly developed commercial and residential areas of the Consolidated Drain watershed. The Consolidated Drain is the single largest discharge of largely untreated stormwater to Portage Creek, a coldwater tributary to the Kalamazoo River in southwestern Lower Michigan. This research effort monitored, evaluated, and identified traditional and innovative stormwater management practices to moderate temperature impacts associated with urban stormwater runoff discharged to Portage Creek via the Consolidated Drain.

The hypothesis was that stormwater systems can be designed to eliminate major alterations to the stream ecology due to thermal enrichment through enhancement or substitution of traditional design and BMPs. This hypothesis was tested through monitoring and evaluation of the hydrologic and thermal regimes of the Consolidated Drain system and the receiving stream, Portage Creek, both prior to and after installation of the treatment system (i.e., the pre-construction and the post-construction monitoring periods).

The three-cell, off-line stormwater treatment system directs runoff through a series of treatment cells. These consist of a 1.01-ha sediment forebay, a 4.65-ha wet detention pond (the two traditional open detention ponds), and a 4.45-ha wetland treatment cell (an enhanced natural wetland). Treated stormwater is then discharged at a moderated rate to a reconstructed

Consolidated Drain channel and ultimately to Portage Creek. The treatment system was constructed and fully operational beginning in late July of 2001.

Continuous monitoring of the hydrologic and thermal conditions of the drain started in the spring of 1999 and continued through September 2003. Water temperature, rainfall, stream flow, air temperature, relative humidity, and incoming solar radiation were all measured at various locations throughout the study area using sensitive recording devices. These data provided valuable insight into how alterations to the drain affect the temperature of the water reaching Portage Creek.

The three objectives of this research included:

- Confirming the relationships between disrupted heat balances found in urban areas and local hydrology
- Quantifying the impacts of thermally enriched stormwater discharges on coldwater stream temperature regimes as related to their influence on aquatic life and habitat
- Identifying effective management and design practices for stormwater BMPs and controls necessary to mitigate problems associated with urban stormwater thermal enrichment

One of the key features of this project was the development of transferable evaluation protocols, stormwater BMPs, and assessment tools for other locations where the thermal impacts of urbanization continue to encroach upon and impair aquatic habitats.

Evidence from the pavement and runoff temperature monitoring for the Crossroads Mall sub-watershed in the Consolidated Drain watershed revealed that pavement surfaces absorbed, stored, and released heat, causing runoff temperature increases. Pavement surface temperatures were regularly measured as much as 22°C higher than air temperatures during summer months. Conversely, tree-shaded pavement surface temperatures were up to 17.6°C cooler than unshaded areas on sunny days.

Analysis of monitoring data showed temperature increases in the two traditional open detention ponds (forebay and wet detention pond) and the ability of a well-shaded wetland to reduce the elevated temperature of open pond outflow to near that of ambient air. During a post-construction modeling period (July 29, 2002, through September 8, 2002), comparing the daily average temperature of air with the treatment cells revealed that stormwater detained in the two open ponds was always warmer (up to 6.4°C) than the ambient air. In contrast, the wetland was able to mitigate this thermal enrichment and maintain wetland outflow temperatures close to the ambient air (ΔT average 0.1°C). A dense vegetative cover blocked direct solar radiation from reaching the standing water in the wetland and enabled it to equilibrate with the ambient air temperature. However, because Portage Creek almost always had a lower daily temperature than ambient air (ΔT average -3.2°C), the treatment system outflow would still have had a warming effect on the receiving stream had it drained directly into the creek. A groundwater fed baseflow in the reconstructed drain channel and from an adjacent drain offset this potential warming as it mixed with treatment system outflow.

Eleven individual storm events in 2002 were separated from the post-construction modeling period to study the treatment system's effect on stormwater thermal loading by water mass balance modeling of the system. Los of flow via evaporation, including evapotranspiration, on average accounted for 19.9% of the total flow reduction by the treatment system while infiltration represented 40.8% of total flow reduction. Though both of these flow loss mechanisms were important factors in the water balance of the system, infiltration was clearly more dominant. Most of the infiltration took place in the wetland. The forebay-wet pond combination had a mere 3.2% flow reduction compared to the wetland's 44.0% flow reduction.

Due to the exposure of detained water in the forebay and wet pond to solar radiation, these two cells combined had small or even negative heat reduction for the 11 events, with an average value of -0.9%. In contrast, the wetland had a higher average heat load reduction (49.8%) than the average flow reduction (44.0%), suggesting that the shading provided by the dense vegetation in the wetland reduced solar warming of the detained stormwater.

There is a strong correlation between flow reduction and heat reduction for the 11 monitored events, suggesting that the primary heat load reduction mechanism is flow loss via infiltration. Data analysis and linear regression revealed that the system could achieve sizable heat load reductions even when the flow reduction was negative, implying other mechanisms were also involved. Shading provided by vegetation in the wetland is likely the most important element in this regard.

Groundwater temperature regimes are influenced by annual temperature variability and are therefore cooler than surface water regimes which are influenced by variability over shorter time periods during critical months. The only way to return the temperature regime of surface water to that of groundwater is to return stormwater to the groundwater. Thermal pollution cannot be physically removed from the water, and heat exchange is a continuous process that does not stop with the cessation of rainstorms and runoff. Thus, detention ponds may be inappropriate for terminal treatment of urban stormwater as surface water that is returned to coldwater receiving streams.

The concept of temperature equivalent (TE), the flow-weighted average temperature of a flow component in the watershed calculated in a pre-defined period of time, was used to reveal the thermal loading contribution (or mitigation) of various flow components in the watershed. TE was applied to identify the locations and potency of thermal enrichment areas in the comparative time period (August through September) of 2002. Stormwater runoff from the separately storm sewered, 30.8-ha Crossroads Mall sub-watershed had a TE value of 23.3°C, 3.3°C higher than the average air temperature and 5.8°C above the Portage Creek TE. This suggests that surface runoff from the mall was effectively transferring heat stored in pavement and roof surfaces to stormwater runoff.

After a prolonged period of detention time in the sediment forebay and wet detention pond, the TE value of the stormwater rose to 24.1°C, a clear indication of the warming effect of these open ponds. The wetland, in contrast, was able to cool the outflow from the wet pond to a TE of 21.6°C. This TE, however, was still 4.1°C higher than Portage Creek. The other two flow components in the post-construction Consolidated Drain watershed, the Schuring Drain (a tributary to the Consolidated Drain main channel) and the Consolidated Drain channel below its confluence with the Schuring, have a steady, cold groundwater base flow. As a result, the TE value of the Schuring Drain flow (16.2°C) was the lowest among all the flow components, including Portage Creek. This suggested that the Schuring Drain constituted a thermal pollution mitigating source to the wetland outflow. This mitigating effect was particularly important during dry weather conditions where the wetland continued to discharge warm outflow to the Consolidated Drain from storage.

Analysis of monitoring data from temperature gauges located in the Consolidated Drain channel and Portage Creek (both upstream and downstream of its confluence with the Consolidated Drain) illustrated thermal regime changes in Portage Creek induced by stormwater runoff during the comparative time period from 1999 through 2002. Sharp temperature increases in the Consolidated Drain channel and downstream Portage Creek were noted during rain events in the pre-construction years of 1999 and 2000. This indicated the flashy thermal impact of the Consolidated Drain watershed on the creek associated with largely unmitigated stormwater discharges to the drain. In the post-construction year of 2001, although these increases still existed, they were smaller in magnitude, indicating the effectiveness of the treatment system in mitigating the acute thermal impact by stormwater runoff on the receiving stream. However, the treatment system, at its best, reduced stormwater temperatures to near ambient air in the wetland. In 2001, this supplied a steady, although small in magnitude, source of heat to Portage Creek. This suggests that chronic temperature impacts (slightly higher than average summer temperatures in Portage Creek) resulted from the regional treatment system. Subsequent 2002 and 2003 data indicate that growth of riparian vegetation in and around drain channels has largely mitigated this chronic effect.

Analysis of daily temperature fluctuations in the system indicated that although such fluctuations were primarily a result of ambient air temperature changes, they can be exacerbated by storm events. The highest single daily temperature fluctuations (daily maximum minus daily minimum) for the Consolidated Drain (9.6°C) and Portage Creek (7.2°C) during the two month comparative period in 2000 and 2001, were both recorded on September 8, 2001. This corresponded to a one-hour 17.8-mm mid-afternoon rain event. The Consolidated Drain contributed to the daily fluctuation in Portage Creek by 2.4°C on the same day. During dry weather, however, vegetative cover played a significant role in regulating daily stream temperature variations in the Consolidated Drain.

A mechanistic model framework was developed for thermal enrichment of urban runoff. Programmed in MatLab language, the HEATRAN (heat transfer) model was developed and completed for the Crossroads Mall sub-watershed. This model was developed to compute runoff temperature directly to evaluate the physical mechanisms that drive thermal enrichment of runoff. HEATRAN uses a distributed heat balance for generalized components of a drainage area. The central idea of the model is the combined heat balance and mass balance. HEATRAN discretizes a watershed into four categories of hydraulic elements and connects them as a network. These four categories are: 1) overland flow; 2) closed conduit or pipe flow; 3) open-channel flow; and 4) open storage (ponds and wetlands). The model covers principal heat transfer processes for an open water body: convection, rainfall mixing, evaporation (sensible heat), radiation, and conduction. HEATRAN is applicable to an infinitely large and complex drainage network.

Model calibration using monitoring data from two afternoon storm events for the Crossroads Mall subwatershed demonstrated that HEATRAN is reasonably efficient and computes temperature changes in a comprehensive way for a typical urban catchment. Model analysis and sensitivity tests identified several factors that influence runoff temperature, including:

- Rainfall characteristics (e.g., duration, amount, and temperature) and temperature difference between rainfall (or air) and the ground surface are the two most important controlling factors for the heat transfer process during wet periods where the effect of radiation and evaporation is limited.

- Rainfall intensity and runoff flow depth increases runoff temperature.
- Pipe networks have a cooling or negative thermal enrichment effect.

Fish, benthic macroinvertebrates and periphyton were sampled at six locations in the Consolidated Drain, Schuring Drain, and Portage Creek during portions of the five-year study period. A regionally specific relationship between stream thermal regimes and fish community structure in the Lower Peninsula of Michigan was applied to the fish data analysis as related to the impact of the stormwater treatment system. Based on its thermal regime, Portage Creek can be classified as a coolwater-moderate stream at sites upstream and downstream of the Consolidated Drain in both the pre- and post-construction time periods. The post-construction downstream assemblage more closely resembles the upstream with a shift toward greater dominance in coldwater guild species, species occupying similar ecological positions within the same habitat, and lesser contribution of coolwater guild species. This suggests improvement in the downstream thermal regime associated with the 2001 completion of the treatment system. Long-term sampling will be necessary to establish any conclusive relationship capable of separating the effects of thermal regime improvements over other stressors associated with the urban setting.

Macroinvertebrate and periphyton metrics measured under pre- and post-construction conditions in Portage Creek did not show obvious changes that can be associated with the construction of the stormwater treatment system. Nevertheless, the Consolidated Drain channel developed a healthier post-construction macroinvertebrate community with the appearance of caddisfly species and a more balanced community. In addition, periphyton community dominance shifted from warm (blue-greens) to coldwater (diatoms) species after the construction. The Schuring Drain channel underwent a similar periphyton community shift.

Overall, this study has identified the mechanisms that cause thermal enrichment of urban stormwater runoff as well as the BMPs and approaches necessary to mitigate this thermal enrichment of receiving streams. In natural or pre-development conditions, infiltration mitigates the impact of runoff and reduces pollutant and thermal loads to receiving water bodies. Runoff can gain heat from impervious surfaces. Shading of pavement or use of highly reflective surface materials provides some benefits in limiting heat absorption, but is somewhat limited by practicality in traditional urban landscapes. Conventional detention ponds achieve hydraulic mitigation and pollutant reduction by collecting and detaining runoff but bypass groundwater recharge in lieu of a direct surface water discharge. Thermal enrichment results with the exposure of detained stormwater to incident solar radiation in such ponds. Thus, alternative or modified strategies to mitigate thermal enrichment associated with traditional stormwater treatment approaches must be instituted to protect the temperature regime of coldwater receiving streams. This can best be accomplished through reduction of return volumes of treated stormwater through infiltration and evapotranspiration (the former is more effective). Shaded wetland treatment cells can achieve targeted detention times for pollutant removal as well as attain ambient air temperatures. Shading of discharge or other treatment system channels minimizes solar heating and protects the thermal regime of groundwater baseflows. Ultimately, the study lends support to general trends in stormwater engineering that target source control solutions with infiltration and recharge, thereby eliminating thermal enrichment concerns associated with surface runoff into coldwater streams.

CHAPTER 1.0

INTRODUCTION

This chapter provides the background and impetus for examining thermal enrichment of stormwater in an urban setting. The hypothesis of this research, the approach to testing this hypothesis, study participants, and overall objectives of this effort are presented. The chapter concludes with the organizational structure of this report.

1.1 Background

Substantial efforts will be undertaken over the next several decades to implement Phase II stormwater regulations, total maximum daily loads (TMDLs), voluntary watershed management plans, and habitat improvement programs. To meet watershed preservation and improvement goals, resource managers often implement a combination of structural and non-structural best management practices (BMPs) to mitigate or correct disrupted stormwater flow and constituent loading in urban areas. Monitoring programs typically only quantify these two effects following implementation. Thermal enrichment of coldwater streams by urban stormwater runoff is often overlooked, and enrichment can be exacerbated by traditional BMPs (Galli, 1990).

Coldwater streams in relatively undisturbed catchments typically respond to storm and runoff events with a decrease in temperature due to cooler climatic conditions and an increase in groundwater contributions. Urban conveyance systems rapidly deliver surface runoff to receiving waters. Since coldwater streams are often driven by groundwater inputs, the delivery of surface runoff at any temperature above that of the receiving stream could be defined as thermal loading. Impervious ground cover is linked to altered flow and temperature regimes in urbanizing areas. Extensive impervious ground cover increases storm runoff volume, changes time and duration of delivery, and stores energy from solar radiation. Thermal enrichment of surface runoff occurs when: 1) stored energy is transferred from impervious surfaces to runoff; and/or 2) temporarily detained stormwater gains energy in conveyance channels and impoundments (e.g., wet detention ponds).

The key to evaluating alternatives to mitigate or prevent thermal enrichment requires recognition and weighting of the elements driving thermal loading in a watershed. Ultimately, runoff volume and timing (driven by impervious cover) and energy exchange between runoff, the atmosphere, and contact surfaces determine the magnitude of stormwater thermal loading to receiving streams. Changes to runoff volume, timing, and energy content occur in four generalized components of typical urban runoff conveyance systems: 1) overland flow; 2) pipe conveyance; 3) open-channel conveyance; and 4) ponds, wetlands, and other waterbodies. An understanding of the influence of each component on runoff volume, timing, and energy can lead to the formulation of mitigating strategies.

Initiated in 1999, this study undertook to understand and quantify thermal enrichment associated with urban stormwater runoff to coldwater receiving streams. Through detailed field research and conceptual and mechanistic model development, approaches to moderate temperature impacts associated with urban stormwater runoff were examined. Monitoring data were collected in an ideal urban setting before and after construction of the Regional Stormwater Quality and Trailways Facility, an urban drain improvement project (completed in July of 2001) in the Consolidated Drain, Portage, Michigan. This system is comprised of three BMPs arranged in series, including a sediment forebay, wet detention pond, and treatment wetlands. The relative effects of each system component on runoff temperature were examined in relation to hydraulic and climatic conditions ultimately affecting Portage Creek, a coldwater receiving stream.

1.2 Project Objectives

Processes that drive stormwater thermal enrichment were investigated to determine strategies that mitigate its impacts on the thermal regime of coldwater receiving streams. The hypothesis was that stormwater management plans and treatment system designs could be developed to significantly reduce or eliminate thermal enrichment through enhancement or substitution of traditional designs and BMPs. This hypothesis was directly tested by monitoring and evaluating the hydrologic and thermal regimes of an existing urban drain and its receiving stream before and after the installation of the offline regional stormwater treatment system. This effort included year-round field monitoring with a specific focus on the warm season months of May through September as related to critical thermal tolerance limits of coldwater stream biota.

Three key objectives of this research were necessary for addressing short- and long-term design needs and habitat protection:

1. Confirming the relationships between disrupted heat balances found in urban areas and local hydrology;
2. Quantifying the impacts of thermally enriched stormwater discharges on cold water stream temperature regimes as related to their influence on aquatic life and habitat; and
3. Identifying effective management and design practices for stormwater BMPs and controls necessary to mitigate problems associated with urban stormwater thermal enrichment.

The study also targeted examination of the likely impacts of thermally enriched stormwater on aquatic life. The approach was to characterize biota present before and after construction of the treatment system, assuming the thermal regime in the Consolidated Drain and the receiving stream would change after construction.

1.3 Participants

The project team was comprised of Kieser & Associates (K&A), the research group of Dr. William James of the University of Guelph, Ontario, Canada, and representatives of URS

Corporation. Project administration and technical management was undertaken by K&A as the prime contractor; Mark S. Kieser was the principal investigator (PI) and project director.

Roles and responsibilities of the project team under the direction of Mark Kieser included:

- URS Corporation addressed engineering controls along with BMP design review of the monitored stormwater treatment system.
- Dr. William James, P.E., acted in the capacity of science advisor. Dr. James and his students also developed a pavement temperature heat budget model to investigate heat exchange between impervious surfaces and surface runoff.
- Dr. Ann St. Amand of Phycotech quantified periphyton and macroinvertebrates.
- Key personnel at K&A included Michael Crimmins (meteorology, geographic information systems, database management), Brian Boyer (field investigations, quality assurance/quality control [QA/QC]), Patty Hoch-Melluish (biological monitoring, BMP review), Dr. "Andrew" Feng Fang (thermal analysis), Nicole Ott (biological monitoring, field investigations), Jeffrey Spoelstra (field investigations, database management), Kevin Ziemer (field investigations), and Joshua Kieser (pavement temperature studies).

1.4 Organization of the Report

The following chapters describe the approach, methodology, results, and conclusions of this study.

- Chapter 2.0: Background
- Chapter 3.0: Site Description and Study Methods
- Chapter 4.0: Temperature Exchange with Contact Surfaces and within Conveyances
- Chapter 5.0: Treatment System Effects on Stormwater Thermal Loading
- Chapter 6.0: Stormwater Thermal Impacts on Portage Creek
- Chapter 7.0: Modeling of Stormwater Thermal Enrichment at the Crossroads Mall
- Chapter 8.0: Conclusions and Implications
- Appendix A: Monitoring of the Consolidated Drain and the City of Portage Stormwater Quality and Trailways Facility
- Appendix B: Instrumentation and Database Coding
- Appendix C: Bioassessment
- Appendix D: Model Background Theory
- Appendix E: 2003 Data Analysis
- References

CHAPTER 2.0

BACKGROUND

Coldwater streams are unique and sensitive features in the landscape. This chapter details the impacts of urbanization on the thermal regime of coldwater streams.

2.1 Coldwater Stream Habitats

The temperature of coldwater streams in Michigan is mediated by groundwater input and preserved by riparian shading. Both of these drivers are altered by urbanization which results in: 1) decreased stormwater infiltration lowering groundwater contributions to streams; 2) increased surface runoff at or above ambient air temperatures; 3) increased connected impervious surfaces as sources for heat exchange with runoff; and 4) increased runoff volume held in surface conveyances and BMP impoundments. All these create the potential to degrade the sensitive thermal regime of coldwater streams. In Michigan, groundwater temperature is typically within 1°C of the local mean annual air temperature (10°C to 13°C), and streams maintain cold temperatures even during the warmest summer months (Collins, 1925; MSUE, 1999). Shading by riparian vegetation preserves coldwater stream temperature by providing insulation from the dominating effects of solar energy input (LeBlanc et al., 1997). Therefore, coldwater streams are typically restricted to headwater areas where channels are narrow enough that dense canopy exists (Vannote et al., 1980).

While it is established that instream management activities (e.g., reservoirs) alter thermal regimes (Ward and Stanford, 1979), related research has only begun to demonstrate terrestrial links between urbanization and stream thermal enrichment (Schueler, 1987; Galli, 1990; James and Verspagen, 1997; Van Buren et al., 2000a). Furthermore, inadequacies of current urban BMPs have brought thermal enrichment to the forefront of stormwater management (Maxted and Shaver, 1999).

2.2 Thermal Loading Changes Due to Urbanization

Watershed urbanization typically leads to increased impervious surface cover (e.g., parking lots, roads, roofs). Runoff characteristics of impervious surfaces alter stream flow regimes and require the installation of stormwater conveyance systems. Together, altered volume and timing of stormwater runoff contribute to acute and chronic thermal loading of receiving streams.

2.2.1 Flow Regime Change

Coldwater streams are typically found in headwater areas and are hydrologically stable due to steady groundwater sources (Vannote et al., 1980). Impervious surface cover and stormwater management practices change the flow source of coldwater streams from groundwater to a mix of groundwater and surface water (Booth and Jackson, 1997). Urbanization increases impervious ground cover, resulting in less infiltration and an acute flow response when stormwater rapidly runs off hard surfaces (Dunne and Leopold, 1978). Stormwater management has traditionally targeted this peak flow response and its associated pollutant load with BMPs that impound water temporarily before releasing it at a more natural rate in an attempt to mimic the natural flow regime. However, there is growing evidence that these BMPs do not adequately compensate for the increased flow volume associated with urbanization, and impacts on receiving waters are unavoidable (Booth et al., 2002; Booth and Jackson, 1997; Maxted and Shaver, 1999). As the hydrologic regime changes from natural to managed, changes occur in the thermal regime of the receiving stream. These thermal regime changes may help explain continued biotic degradation of receiving waters despite BMP implementation.

2.2.2 Acute Thermal Loading

An acute thermal loading condition exists when runoff is conveyed rapidly to the receiving stream. This acute condition may induce a thermal shock as stormwater at or above ambient air temperature reaches the coldwater receiving stream, which by definition, has a lower water temperature than the ambient air. Thermal enrichment of runoff occurs when runoff gains additional heat due to heat exchange with impervious surfaces during overland flow. Temperature changes of this nature are abrupt but streams recover relatively quickly.

Urbanization increases solar energy absorption and storage by built surfaces (e.g., buildings, concrete, roadways), thereby amplifying annual temperature extremes (Goward, 1981). This urban heat island effect (Taha, 1997) drives summer temperatures in cities above temperatures in surrounding rural areas. Materials used in urban land covers, such as stone, asphalt and concrete, do not reflect light very well and effectively absorb incoming shortwave solar radiation. As a result, surface radiant temperatures exceeding 70°C can be measured on low reflective surfaces, such as asphalt and concrete, during summer months (Kieser et al., 1999). Hot surfaces emit longwave thermal energy. This causes a perceptible heat rise in the immediate air layer above these low albedo surfaces. Since incoming solar radiation is the primary driver of urban heating, the urban heat island effect is most prominent in summer months. Average daily summer air temperatures are now 2.5°C higher in some urban areas than in surrounding rural areas (Bretz et al., 1998).

Because of the high thermal storage capacity of urban land cover materials, the flux of surface longwave thermal energy from these surfaces can lag into the evening when direct solar forcing is absent. This longwave thermal energy flux from urban surfaces contributes to higher ambient air temperatures through the overnight hours. Doll et al. (1985) showed that concrete and asphalt slabs, which have a similar albedo to soil, have a higher thermal storage capacity than bare soil alone, thus illustrating that even materials of similar albedos can have very different energy budgets over a diurnal cycle.

The overall effect of urbanization on stream hydrology and thermal regime is amplified by the timing and nature of warm season precipitation events. In a climatological study of lightning data, Crimmins (1998) showed that southern Michigan, for example, experienced most thunderstorms between 4 p.m. and 7 p.m. Convective thunderstorms are the heaviest and shortest duration precipitation events. They can produce large volumes of runoff that need to be conveyed as stormwater in urban watersheds. These storms occur when pavement temperatures are likely at their hottest, thereby increasing acute heat load in this Midwestern United States.

2.2.3 Chronic Thermal Loading In Conveyance and Treatment Systems

Chronic warming is viewed as the long-term thermal condition reflected by average daily, weekly, or monthly temperatures in the critical summer months. Differences of up to 11°C in daily temperature range have been noted between forested and agricultural streams (Klein, 1979). Similarly, urbanization can raise average stream temperature as much as 5 °C to 8°C (Pluhowski, 1970). Chronic warming occurs in urbanizing streams and channels because reduced baseflow is influenced by ambient air temperatures and direct solar radiation. Riparian vegetation provides shade from direct sunlight and reduces the direct absorption of solar radiation. Ambient air temperatures are also balanced by vegetation through evapotranspiration and reduction of wind velocities. Galli (1990) illustrated how removal of riparian vegetation can raise stream temperatures 6 °C to 11°C in the summer and reduce stream temperatures by up to 4°C in the winter in Maryland streams. These thermal regime changes alone are enough to change aquatic ecosystem structures. Urbanization tends to result in less riparian canopy cover, exacerbating the warming effect of urbanization on streams and channels.

Chronic warming also occurs in ponds and wetlands. Schueler (1987) explained that temperature increases of runoff could occur: 1) between runoff and impervious urban surfaces; 2) in unshaded conveyance channels; and 3) in impoundments such as stormwater detention ponds. Galli (1990), working in Maryland, demonstrated all three conditions in urbanized watersheds compared to an undeveloped reference watershed. He showed: 1) higher stream temperatures directly related to increasing levels of percent impervious surface across watersheds; 2) increased temperature of runoff as it passed through open channels; and 3) higher outflow temperatures (compared to inflow) in online stormwater management detention structures. In another study of the thermal balance of an online stormwater treatment pond in Ontario, Canada, Van Buren et al. (2000a) also demonstrated net energy gain (thermal enrichment) in the pond during dry weather conditions. BMPs mitigate flow and treat for common pollutants but create a potential thermal loading condition.

A common design suggestion for stormwater ponds and wetlands is a treatment train approach and maintenance of shading to minimize solar thermal input. Since pond temperature regimes in small, shallow, shaded water bodies closely track that of air temperature (Kadlec and Knight, 1996), fully-shaded urban BMP detention systems will also approach an air temperature balance. Therefore, treated stormwater release from fully shaded ponds will be warmer than coldwater receiving streams.

The tendency for impoundment BMPs to hold water for an extended period will lead to surface water held in balance with surface temperature drivers. There is much concern that impoundment BMPs do not adequately mimic natural flow regimes, and that limitations or improper design leads to an increase in sediment moving flow (Roesner et al., 2003; Booth et al., 2002). The implication is that to mitigate this condition, water must be released slower to mimic runoff duration, leading to longer detention time and acclimation of runoff with surface water temperature drivers.

2.2.4 Acute Loading Due to BMPs

Stormwater treatment systems have the potential to increase thermal loading in an acute fashion when detained water from previous storms is forced out of holding ponds with the first flush from a new storm. Typical releases in stormwater ponds are from surface layers that have been exposed to solar heating. Few such ponds release water at depth due to pond size limitations and water quality concerns (Galli, 1990). Van Buren et al. (2000a) demonstrated net loss of heat from an online detention pond during wet weather as plug flow pushed previously impounded warm water out of the pond and replaced it with cooler runoff from the watershed.

2.3 Urban Thermal Energy as a Pollutant

In order to manage urban stormwater thermal loading it is critical to understand the difference between thermal energy and more traditional pollutants (e.g., phosphorus and sediments). The following sections introduce the challenges of treating thermal energy as a more traditional pollutant.

2.3.1 Critical Periods

Midwestern coldwater streams are generally most sensitive to thermal degradation during summer low flow periods. Galli (1990) defined critical periods for protection of coldwater streams in Maryland as occurring from late May to late September, the periods of high input of solar energy to impervious surfaces and stormwater management ponds. LeBlanc et al. (1997) more generally defined critical time periods as those occurring during summer months when stream thermal capacity was low due to reduced volume. Hinz and Wiley (1997) indicated that temperatures of coldwater streams in their study region (overlapping this study) approach levels critical to biota during July.

2.3.2 Monitoring

Unlike common pollutant mass loading (e.g., total suspended solids [TSS] and total phosphorus expressed in pounds), thermal energy as a pollutant is in a constant state of flux on all time scales. Limitations to temperature monitoring associated with urbanization include the necessity for long-term study of the thermal complexity of watersheds and the synergistic effects of the physical and hydrologic changes to receiving waters that occur due to urbanization. Despite these challenges, temperature is an easily measured, useful indicator with which comparisons can be readily made 1) between urban runoff and receiving waters; 2) between upstream and downstream waters; 3) between natural and developed watersheds; 4) in stormwater conveyances and ponds; and 5) to identify potential species loss (Program Managers: EOA Inc., 2001; Claytor and Brown, 1996).

In the National Pollutant Removal Performance Database (Winer, 2000), no temperature information is available for the various BMPs reported. In the National Stormwater Best Management Practices Database (2002), the lack of detailed data collected for the purpose of assessing thermal impacts is also notable. Only a limited number of studies have monitored in detail the thermal impacts associated with urbanization and the effects of traditional stormwater

BMPs (e.g., Galli, 1990). More typically, measurements of water temperature were limited to discrete measurements when grab samples were collected with field parameters (e.g., pH, dissolved oxygen, and temperature). Rigorous analysis of such data was often outside the purposes of monitoring efforts. Moreover, calls for standardization of BMP evaluations do not address thermal impacts (Strecker et al., 2001, 2002), nor do current detention design revisions consider the temperature of treated stormwater. Instead they focus on flow mitigation and removal of common pollutants (Roesner et al., 2003).

2.3.3 Stormwater Management Inertia

Although thermal impacts have been recognized as an important issue for urban stormwater design considerations (Schueler and Galli, 1995), in practice engineering designs typically focus only on flow moderation to minimize erosive conditions and reduction of sediment/pollutant loadings. They typically do not incorporate temperature considerations even in coldwater receiving streams (e.g., Booth and Jackson, 1997).

One of the least defined elements of published stormwater BMP studies appears to be design criteria used to engineer and construct treatment systems. These design criteria require additional examination from the perspective of thermal enrichment. Ignoring thermal enrichment of stormwater in BMP design may result in a false sense of accomplishment and allow substantial detrimental impacts to accrue in receiving waters.

Many regionally specific stormwater planning documents acknowledge the potential for certain BMPs to have negative downstream thermal impacts. Most documents reference the work by Galli (1990) when suggesting particular BMP installations to avoid *if* the design is incorporating concerns for thermally sensitive habitats downstream. However, such language is not prohibitive with respect to coldwater streams. The U.S. Environmental Protection Agency (U.S. EPA) calls for maintaining pre-development temperatures or temperatures that sustain downstream biota. The agency does use prohibitive language in coldwater situations, but suggests that shaded ponds will mitigate temperature (e.g., U.S. EPA draft NPS Guide, 2003).

While always recognized as the ultimate solution to urban runoff, use of proactive planning associated with watershed management is finally gaining awareness and momentum within U.S. EPA. Often, this involves the move away from structural BMPs toward non-structural methods and ordinances aimed at increasing onsite infiltration of stormwater. Moreover, the hundreds of temperature TMDLs pending throughout many U.S. regions will explicitly require consideration of mitigation and restoration strategies that focus on the complex yet rarely defined issue of thermal enrichment.

2.4 Predictive Tools

A host of models define the thermal energy balance driving water temperature, but not all lend themselves to site-specific management. Standard models use a heat budgeting approach to predict stream temperature changes. Emerging efforts focus on models that use a terrestrial heat budgeting approach to predict runoff temperature by modeling the thermal balance of generalized stormwater system components. Site-specific decision making regarding runoff management requires models capable of simulating runoff temperature treated or untreated by a possible range of BMPs. Hence, the development of mechanistic runoff temperature prediction models will greatly benefit decision making related to urban runoff temperature management.

Chen et al. (1996) modified the Hydrologic Simulation Program–Fortran (HSPF) to simulate stream temperature in Oregon by incorporating a module for riparian shading and applying it to multiple stream reaches across a watershed. The model demonstrated the importance of shading and was useful in decision making for the nation's first temperature TMDL for the Upper Grande Ronde Watershed.

While useful for natural stream temperature prediction, established models lack site-specific urban criteria. With the Critical Urban Stream Temperature Model (CrUSTe), LeBlanc et al. (1997) showed that three in or near stream parameters affect stream temperature in unregulated natural streams including: 1) riparian vegetation; 2) low flow regime; and 3) channel morphology. These drivers are highly modified in urbanizing watersheds and are suggested to predict the effects of urbanization. However, criteria for managing thermal impacts of stormwater runoff before it reaches the receiving stream are still lacking. CrUSTe does not model the direct effect of heat buildup and washoff. Similarly, common stream temperature models (e.g., HSPF and Instream Water Temperature Model [SNTMP]) have limited ability to model stormwater temperature and no ability to consider heat washed off impervious surfaces (Johanson et al., 1984; Theurer et al., 1984; James and Xie, 1999).

The Thermal Urban Runoff Model (TURM) has been developed for the Dane County, Wisconsin, stormwater management program and is used as a site-specific planning tool (Roa-Espinosa et al., 2002). It is required that proposed land-use changes result in cooler runoff temperature unless it can be demonstrated using TURM that the runoff temperature increase will be zero. TURM accounts for heat absorbed by runoff from impervious surfaces, and the Web interface allows the user to consider the implementation of a stone crib BMP. TURM is in its introductory stages and unavailable for critical evaluation.

Heat budgeting models that simulate stored energy transfer between impervious surfaces and stormwater runoff have been developed and refined (James and Verspagen, 1997; Picksley and Deletic, 1999; Van Buren et al., 2000b; Haq, 2001). Associated with this Water Environment Research Foundation (WERF) project, and building off of a decade of work on the subject, Li and James (2003) have developed a conceptual framework and a heat transfer (HEATRANsfer) model targeted for incorporation into the PC Stormwater Management model (PCSWMM) platform that will track heat exchanges that impact runoff temperature in any or all of four generalized runoff components in a watershed: 1) overland flow; 2) pipe conveyance; 3) channel conveyance; and 4) ponds and wetlands. This thermal budgeting model has great potential as a planning tool for BMP comparison and selection when tailored for targeted watersheds.

2.5 Temperature and Stream Biota

Water temperature is the driving force for the distribution and physiology of aquatic biota. It has come into focus with current interest in climate change science (Magnuson and DeStasio, 1997) and Clean Water Act enforcement within the context of temperature TMDLs. Acute thermal impacts to aquatic biota include mortality as well as altered development, growth, and reproduction. Long-term thermal effects associated with chronic loading include evolutionary, physiological, and behavioral responses of fish (Matthews, 1987), invertebrates (Vannote and Sweeney, 1980), and periphyton (DeNicola, 1996).

In Maryland, most coldwater organisms are severely stressed at temperatures above 21°C with a 2-3°C change in temperature enough to eliminate sensitive insect species (Galli, 1990). Both of these temperature thresholds are commonly observed in urban watersheds (James and Xie, 1999). Hinz and Wiley (1997), for example, found simple linear relationships for growth rate and standing stock of juvenile brook and brown trout with mean daily temperature fluctuation in July in streams in the northern portion of the lower Michigan peninsula. Temperature was identified as a master variable with respect to growth and production of fish due to its influence on rates of metabolism and foraging activities.

Traditional laboratory determination of aquatic temperature limits of stream biota focuses on the effects of maximum or average temperature. However, it has been suggested more recently that the range of daily variation is also important and that human activity can influence such daily variations (Matthews, 1987; Wehrly, 1999; Schrank et al., 2003; and Johnstone and Rahel, 2003).

A recent review of urban stream ecology by Paul and Meyer (2001) covered many aspects of the effects of urbanization on various trophic levels. It revealed the scope of the problem of examining the isolated effects of temperature on aquatic biota when urbanization causes a host of stressors to stream health. In most cases, the move toward holistic water quality monitoring involves multimetric biotic indices to track long-term aquatic health. The nature of these indices is such that isolating short-term impacts associated with individual stressors (e.g., temperature) is not practical.

Wehrly et al. (2003) developed a regionally specific relationship between stream thermal regimes and aquatic fish community structure in the lower peninsula of Michigan. By identifying mean temperature categories (cold, <19°C; cool, 19-22°C; and warm, ≥22°C), and three maximum temperature fluctuation categories (stable, <5°C; moderate, 5-10°C; and extreme, ≥10°C) a 3 × 3 matrix was developed with nine thermal regimes. The effects of urbanization on acute and chronic thermal loading can be viewed in light of expected changes in stream fish communities, as it is recognized that the thermal regimes of Michigan streams are highly influenced by catchment and local scale attributes (Wehrly et al., 1998).

Thermal impacts are not necessarily restricted to streams but can also be demonstrated in lakes. Treated urban stormwater runoff discharged to lacustrine environments has been found to exacerbate impacts of thermal enrichment. An in-line wetland constructed in 1985 to treat stormwater inflows to Lake McCarrons in Roseville, Minnesota, was found to increase summer discharge temperatures by 5°C (Metropolitan Council Environmental Services, 1997). These heated inflows reduced the epilimnion of the 50-ha lake, further exacerbating lake water quality conditions despite reduced pollutant levels through the wetlands.

2.6 Importance of this Research

The mechanisms that relate urbanization to the thermal enrichment of stormwater have been recognized. However, the effects of this thermal enrichment on aquatic life and habitat have been largely ignored, despite the fact that temperature is a key habitat parameter for aquatic health. Moreover, traditional stormwater BMP designs do not consider temperature mitigation as a critical design criterion. With millions of dollars to be spent on stormwater mitigation, thermal enrichment must be considered a key component to the long-term success and sustainability of the targeted improvements for receiving waters. Failure to incorporate specific thermal mitigation design criteria into strategies for stormwater management may result in impediments, delays, or failure of the efforts to restore natural flow regimes, reduce stormwater pollutant loads, and protect aquatic habitats. Furthermore, thermal enrichment of receiving waters may contribute to the apparent failure of urban BMPs to protect downstream waters (Maxted and Shaver, 1999).

Managing urban thermal energy as a pollutant requires an understanding of the spatial and temporal heat exchanges that take place as runoff moves over land, through conveyances and ponds, and into receiving streams. Toward that end, this study monitored the thermal regime of an offline, three-stage, regional stormwater treatment system consisting of a sediment forebay, extended wet detention pond, and wetland. This system handles stormwater from a highly urbanized watershed discharging to a coldwater receiving stream. The wetland cell of the treatment system was highly shaded, allowing direct comparisons between unshaded and shaded treatment cells. Changes in thermal conditions (heat load and temperature) were monitored as runoff passed from impervious surfaces, through conveyances and channels, through the cells of the treatment system, and into the receiving stream.

CHAPTER 3.0

SITE DESCRIPTION AND STUDY METHODS

This chapter provides a detailed description of the project setting for thermal enrichment studies and the associated research methods employed.

3.1 Project Setting

This study was conducted in a highly urbanized drainage system with a recently constructed regional stormwater treatment system. The following sections describe the setting of the urban drainage before and after construction of the treatment system.

3.1.1 Urbanizing Conditions

A regional stormwater drain improvement project in Portage, Michigan, focused on a highly impervious urban watershed and its impacts on water quality in a coldwater receiving stream. The Consolidated Drain No. 1 (Consolidated Drain) was an open channel waterway that drained approximately 187 ha in the heart of the city's commercial center, including the regional retail attraction known as Crossroads Mall (Figure 3-1). The drain was constructed by the Kalamazoo County drain commissioner in 1963 and was conveyed to city jurisdiction in 1972. The watershed of the Consolidated Drain has become highly urbanized during the past two decades, and the area's few remaining open spaces continue to be developed. Within this drainage, 60% of the land use is commercial; 33% is residential; and 7% is wooded, wetland, or open space. A second urbanized watershed, the Schuring Drain, discharges into the lower portion of the Consolidated Drain. These 142 ha are comprised of 48% residential; 22% cultivated; and 30% wooded, wetland, or open space.

Other than some limited impoundment in several failing stormwater detention ponds, stormwater runoff from the Consolidated Drain watershed flowed largely untreated in the drain into the receiving stream, Portage Creek. The aquatic habitat of this stream was subjected to the characteristic flashy stormwater discharges and sediment loading, which plague this and other waterbodies throughout Michigan and the United States. In addition to pollutant loading and flashy hydraulics, thermal enrichment by unmoderated urban stormwater discharges additionally altered the ecological integrity and natural character of Portage Creek.

Stormwater loadings from this drain were identified as the largest single contributor of pollutants to Portage Creek in the vicinity of Celery Flats Interpretive Center and Garden Lane, a focal point of the city's trailway and recreational facilities. Portage Creek is classified by the Michigan Department of Environmental Quality as a high quality coldwater stream capable of sustaining trout and other higher order vertebrates (Heaton, 1997). Portage Creek discharges to the Kalamazoo River, which is approximately 12.1 km downstream of the Consolidated Drain discharge. Preserving the natural resources of Portage Creek is an important objective for the quality of life in Portage.

The municipality's goals for the drain improvement project were habitat restoration and protection of the coldwater receiving stream. The City of Portage therefore embarked on a multi-million dollar effort in 1999 to develop the Regional Stormwater Quality and Trailways Facility to address stormwater runoff discharging to Portage Creek via the Consolidated Drain watershed. This project served as a model to address use impairments related to nonpoint source pollution, habitat degradation, and the coupling of these concerns with land use. This project also addressed other pending water quality initiatives in the Kalamazoo River watershed, including Phase II stormwater regulations and the Kalamazoo River watershed/Lake Allegan TMDL for phosphorus (Heaton, 1999). Discharges from the Schuring Drain watershed were targeted for future mitigation efforts as these were typically small in comparison to those from the highly developed Consolidated Drain watershed.

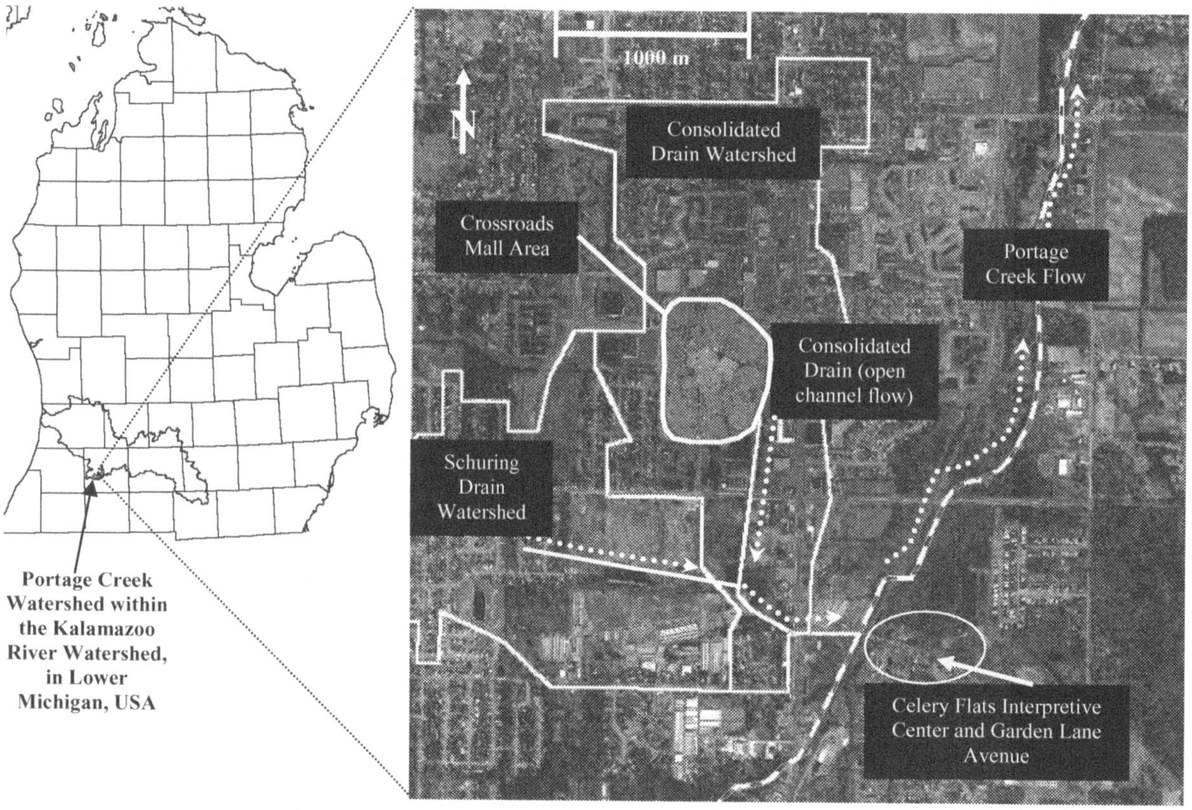

Figure 3-1. Aerial Photo (1997) of the Study Area in Portage, Michigan, Prior to Stormwater Treatment System Construction. The Consolidated Drain and Schuring Drain watersheds are shown. The Crossroads Mall area is visible in the center of the Consolidated Drain watershed.

The city's goals for Portage Creek were accomplished with the design and construction of a regional stormwater treatment system where stormwater is directed from the existing separate stormwater sewer system and open drain into a series of treatment cells (Figure 3-2). These consist of a 1.01-ha sediment forebay, a 4.65-ha wet detention pond, and a 4.45-ha wetland treatment cell. Outflow from the treatment system is discharged at a moderated rate to a reconstructed drain channel and ultimately into Portage Creek. This overall stormwater project incorporated public trailways, a public fountain/ice rink, and educational components and was aptly named the City of Portage Regional Stormwater Quality and Trailways Facility (or stormwater treatment system herein).

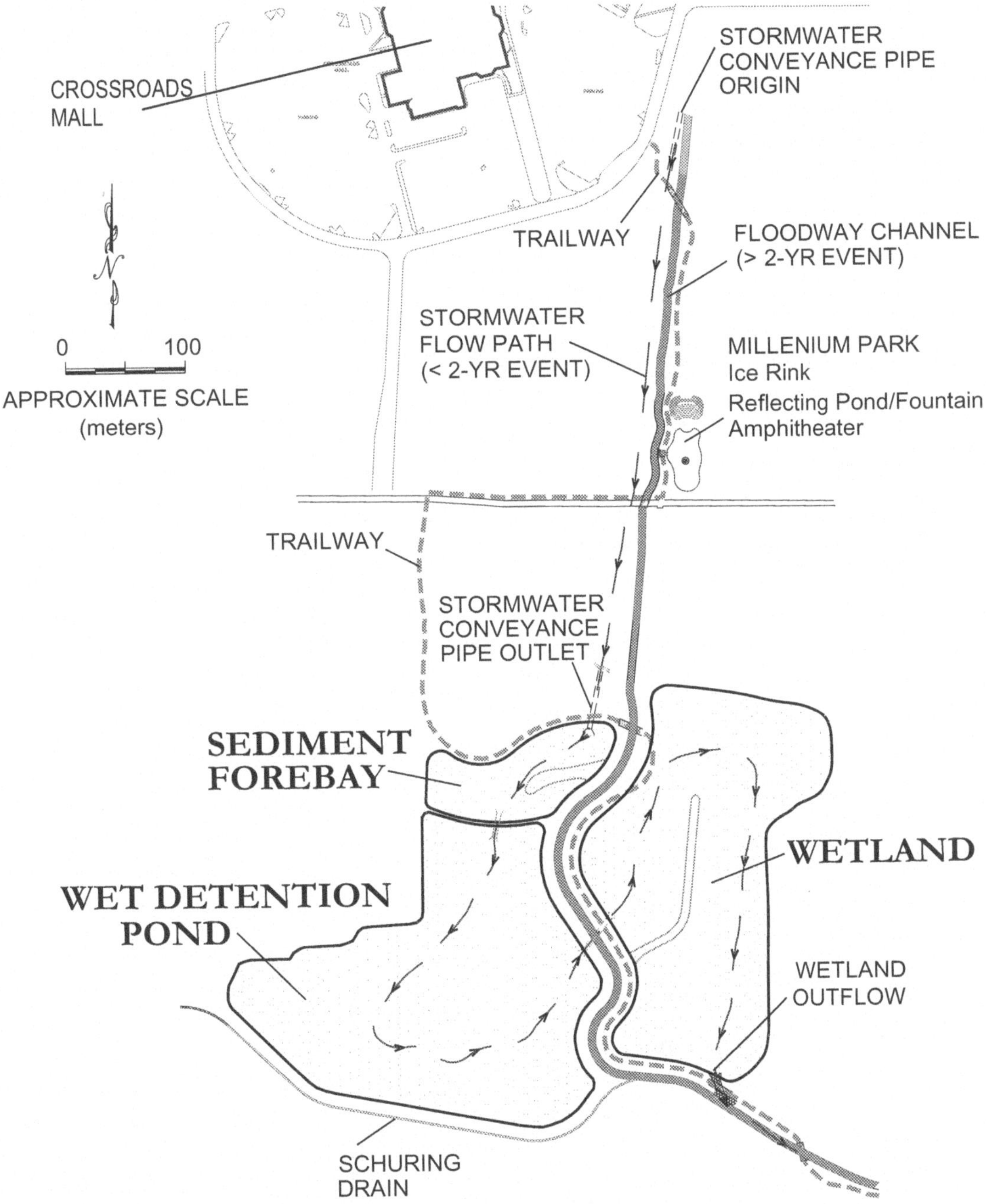

Figure 3-2. Diagram of the City of Portage Regional Stormwater Quality and Trailways Facility.

The environmental setting of this stormwater treatment system provided an ideal field research setting to examine the effects of thermal enrichment on coldwater habitats and to identify new techniques (or modify existing approaches) to mitigate these effects. The enhancements envisioned for the Consolidated Drain were foreseen as a multi-faceted approach to addressing stormwater management needs through water quality and water volume issues. Water quality issues are addressed by combining existing wetlands (the wetland treatment cell and portions of the wet detention pond) with passive biological treatment and filtration methods to mitigate the existing chemical loading prior to discharge into Portage Creek. In order to adequately achieve high levels of pollutant and sediment removal, stormwater passes through a sediment forebay for removal of floatable solids and coarse materials. Stormwater is then detained in the wet pond for approximately 24 hours before it is discharged to the wetland treatment cell for a treatment duration of 5 to 10 days. From the wetland cell, treated stormwater discharges back into the drain channel. The drainage channel has been reconstructed into a more natural watercourse to provide natural habitat, erosion control, and flood protection. The wetland areas not only provide enhanced water cleansing but also excellent natural habitat for a variety of native plants and wildlife.

In order to address stormwater flooding and control issues, stormwater storage in the watershed was also regionalized to eliminate numerous smaller, ineffective detention basins throughout the Consolidated Drain watershed. Regionalization of stormwater in a watershed is predicated upon multiple smaller basins often being designed to release their overflow at the same time during a rain event. Although each basin may function well independently, the net sum of the basins discharging at the same time can lead to flooding and erosion of the drain channel and adjacent property.

A regional method for detaining stormwater eliminated the small, individual detention basins and combined their volume into a larger, more hydraulically efficient area. A large central detention area can be better designed and managed for natural habitat and aesthetics than can multiple individual basins that often become safety hazards and eyesores. Many open stormwater basins were also constructed within the drainage area prior to the understanding that direct stormwater discharge from unlined, single-cell detention basins into the groundwater is inappropriate in this area because of municipal well fields. Retrofitting the numerous smaller basins in the watershed would have been far more expensive and less effective than constructing one central basin. Cost savings for this project were anticipated to be substantial considering regulatory, water quality, habitat, and quality of life issues associated with these improvements.

In summary, the City of Portage Regional Stormwater Quality and Trailways Facility:

- Eliminated several smaller detention areas that were failing or of insufficient capacity;
- Provided improved, multi-celled physical and biological stormwater treatment for the 187-ha Consolidated Drain watershed, which continues to be increasingly developed with high-density commercial facilities;
- Prevented further encroachment by development into the wetland areas of the site, thereby preserving threatened wetlands, open space, and flood storage capacity;
- Improved the water quality of the Consolidated Drain discharge into Portage Creek;
- Restored coldwater instream and bank habitat to better support native biota.

Appendix A documents the extensive pollutant and flow monitoring of the treatment system conducted by K&A in a parallel effort to this study. Relevant results are used for analysis of the thermal enrichment research presented herein.

3.1.2 Portage Creek

The 14,200-ha Portage Creek watershed is rural in its upper reaches and urban in its lower reaches. The drainage includes portions of both the cities of Portage (upper watershed) and Kalamazoo (middle and lower watershed). Portage Creek ranges from a first- to third-order stream and is designated as coldwater habitat in the study area, which is located in the upper reaches of the watershed in the City of Portage (Heaton, 1997).

3.1.3 Consolidated Drain Pre-construction

The Consolidated Drain was an approximately 0.8-km long open-channel conveyance draining 187 ha of largely curb and gutter, municipal separate storm sewer system. The Consolidated Drain watershed dry weather baseflow was approximately 0.0283 m^3/s and consisted of groundwater inputs to the lower reaches of the channel. The Consolidated Drain watershed wet weather flow was composed of: 1) overflow from failing detention basins at the Crossroads Mall property, which has 30.8 ha of 96% impervious surface; 2) uninhibited flow from the Consolidated Drain watershed consisting of 129 ha commercial and residential property; 3) limited contributions from the adjacent Schuring Drain (142 ha), which was largely diverted in a private decorative pond; and 4) minor surface runoff contributions from nearby roads and small parcels. Overall, the pre-construction contribution of runoff to Portage Creek was dominated by flashy, untreated flows from the Consolidated Drain watershed.

3.1.4 Consolidated Drain Post-construction

Separate from WERF-funded elements of the project, the City of Portage Regional Stormwater Quality and Trailways Facility was constructed from August 1999 through July 2001. This off-line passive treatment system began operating in mid-July 2001. The timing of WERF-sponsored funding did not coincide with pre-design or pre-construction phases of the treatment system; however, the authors of this report conducted pre-construction monitoring with other funds. With the completion of the stormwater treatment system, stormwater from the Consolidated Drain watershed (including the Crossroads Mall) is now directed off-line from the existing separate stormwater sewer system into the series of three treatment cells. Treated stormwater that does not infiltrate or evaporate is released through a weir structure from the wetland into the reconstructed Consolidated Drain channel.

In general, the treatment system was constructed to fully capture a two-year, 24-hour storm event. Storms exceeding this return period frequency would result in overflow from weir structures associated with the 30.8-ha Crossroads Mall separate storm sewer system and the other upstream drainage system. First flush and storm volumes up to the two-year storm are still captured while the tails of these larger storms or peak flows may overtop weirs and flow to Portage Creek via the reconstructed Consolidated Drain floodway/drain channel (Figure 3-2). As part of the public amenities of the Regional Stormwater Quality and Trailways Facility, a reflecting pond and groundwater-fed fountain were constructed at the headwaters of the reconstructed drain channel. The reflecting pond doubles as a refrigerated outdoor ice rink in the winter.

As a result of treatment system capacity limitations in terms of storage volume and limited topographic relief, the adjacent Schuring Drain was reconfigured, and its baseflow and storm flow now pass untreated into the open Consolidated Drain channel downstream of the treatment system. The effects of stormwater treatment in the Consolidated Drain must therefore be separated from the post-construction contributions of the Schuring Drain to isolate exclusive changes in thermal loading to Portage Creek based on the treatment system.

Post-construction dry weather flow entering Portage Creek is a mix of: 1) Consolidated Drain groundwater baseflow; 2) supplemental flow from the City of Portage reflecting pond fountain (0.014 m^3/s to 0.028 m^3/s) pumped groundwater during daylight hours, spring through fall); 3) release of treated stormwater from the treatment wetland; and 4) Schuring Drain groundwater baseflow.

In addition to the dry weather flow components, the post-construction wet weather flow entering Portage Creek from the Consolidated Drain includes contributions from: 1) stormwater treatment system overflows at the headwall weirs during high volume or high intensity storms; 2) minor surface runoff inputs to the Consolidated Drain channel from nearby roads and small connected parcels; 3) treatment system emergency overflow at the wet pond under major storm events (though this did not occur during the study); and 4) wet weather flow from the Schuring Drain watershed. Although the treatment system was primarily designed to mitigate flashy wet weather flow and pollutant loading up to the two-year storm, efforts were made to minimize thermal loading of stormwater within the treatment system through:

- Preservation of existing shade trees and planting of new shade trees in strategic locations throughout the system to reduce direct solar heating
- Use of existing shrub and wooded wetland vegetation in the wetland treatment cell
- Use of a naturally occurring marl as a high-albedo bottom surface in the wet detention pond
- Baseflow augmentation with groundwater
- Shaded rock outfalls from the wetland

3.2 Methods

The intensity of monitoring varied during the study before and after construction of the stormwater treatment system based on funding availability and needs identified as the research progressed. The following sections differentiate between time periods and parameters recorded for the various goals of the study.

3.2.1 Temporal Setting

Four distinct monitoring periods to address study objectives were used for detailed analyses:

- Pre-construction (1999 to 2000)
- Post-construction (July 21, 2001 to December 31, 2003)
- Post-construction modeling (July 29, 2002 to September 8, 2002)
- August–September comparative period (1999 to 2003)

These descriptive periods are cited directly in the remainder of this report.

The pre-construction period spanned from initial temperature monitoring in June 1999 through September 2000. Monitoring included temperature measurements taken prior to any ground-breaking in August 1999 through a substantial portion of the treatment system construction period. Although significant earth-moving efforts were underway during construction, stormwater from the Consolidated Drain continued to flow unmitigated into Portage Creek as the original drain channel and/or flows were maintained until the system came on-line in mid-July 2001.

The post-construction period is represented by the time frame from when the system came on-line through instrument downloads in late 2003. A variety of information from this broader post-construction time frame is extracted for discussion and analysis throughout the remainder of this document.

The post-construction modeling period involved intensive data collection in 2002 for a late summer period of continuous flow from the wetland outflow. This period represents the time when the outflow was non-existent immediately prior to and following this 42-day span. Data were used to foster a heat budgeting approach to understand thermal loading characteristics.

The August-September comparative period represented a summer timeframe where temperature data were available for these months in each of the four years of monitoring (1999 to 2003). Data from the August-September comparative period were used to compare thermal loading to Portage Creek both pre- and post-treatment system construction.

During the pre-construction period, monitored parameters within the Consolidated Drain watershed included air temperature, solar radiation, water temperature at various points along the Consolidated Drain and receiving stream, rainfall in and around the Consolidated Drain and greater Portage Creek watershed, and flow in upstream and downstream portions of the open drain channel. During the construction period, excavation and construction of the stormwater treatment system made it necessary to remove some thermistors as well as a level logger at the upstream location in the channel.

Budgetary constraints for the treatment system project caused the City of Portage to delay the purchase of treatment cell component monitoring equipment originally targeted for installation in May 2002. Some limited post-construction equipment was added to the study site in May 2002 by K&A, including additional thermistors, an anemometer (wind speed and direction), and a level logger at the wet pond outlet for quantification of water volume moving through the treatment cells. The City of Portage leased the minimum required monitoring equipment in July 2002. At that time, additional changes to the monitoring layout were made to accommodate the reduced selection of equipment. These included a rearrangement of rain gauges, installation of additional thermistors, and installation of level logging equipment to monitor flow and temperature into, through, and out of each of the system components. These final adjustments were planned to facilitate efforts to provide: 1) adequate data to Dr. William James for the development of a thermal model; and 2) information to track runoff temperature and volume into, through, and out of the treatment cells to model the characteristics of the stormwater treatment system using a heat budget approach.

3.2.2 Instrumentation

Appendix B and Figure B-1 provide details of instrumentation locations, characteristics, and period of record for monitoring equipment associated with the project. Instrument names include a two-digit number corresponding to the instrument location.

The most prominent instrumentation used to monitor water temperature was the Standard Range Hobo Tidbit thermistor logging in centigrade at five-minute intervals. Thermistors were housed within or secured to permanent perforated standpipes or structures. Periodic laboratory testing confirmed that thermistor accuracy was within the manufacturer's specifications.

Up to five RainWise tipping-bucket rain gauges with Hobo event loggers and two ISCO rain gauges recorded rainfall in one-hundredth inches per tip. The pre-construction and post-construction gauge networks were maintained by K&A and were arranged within the study area redundantly as well as around the Portage Creek watershed.

Three meteorological instruments were located at Station 12 (Figure B-1) within the wetland treatment cell. A Hobo air unit was mounted 2 m above the ground surface and measured air temperature and humidity at 10-minute data intervals. A Campbell Scientific pyranometer was mounted 3 m above the ground surface and measured incoming total solar radiation (MJ/hr). A Campbell Scientific logger with a Wind Sentry anemometer was mounted 4 m above the ground surface and measured wind speed and direction at 10-minute intervals.

Global Water level loggers, ISCO level loggers, and an ISCO area velocity meter recorded water levels at five- or 10-minute intervals at critical points in the treatment cells and channels during select pre- and post-construction time periods. An elevation/flow relationship was established for each location by: 1) development of a cross-sectional stage discharge curve; 2) fixed structure elevation calculation (e.g., weir); or 3) area velocity calculation (e.g., submerged intermittent pipe flow). Appendices A and B provide additional details.

Surface temperatures were recorded in the Crossroads Mall parking lot with an infrared Raytek Raynger ST thermometer. Quantitative distributed transects and qualitative spot temperatures were conducted at different times under varying weather conditions before, during, and after a limited number of storms.

3.3 Adherence to Standard Operating Procedures

All field instruments were installed, calibrated, and downloaded in accordance with the manufacturer's recommended use by trained K&A personnel. Upon return to the office, data were immediately entered into the project database, which was backed up on a regular schedule.

Field and office notebooks were established to allow for accurate transferability among K&A personnel, maintain scientific integrity, and ensure appropriate QC/QA efforts. Field activities were logged as they were completed.

The office notebook contained standard operating procedures (SOPs), instrument download schedules, and instrument manuals. The SOPs were established to address the nuances that occur with use of the selected equipment. All personnel performing duties in the field or working with the data were versed in the established SOPs. The SOPs were not scientific documents, but rather an aid to personnel to ensure transferability and to maintain a level of consistency among personnel across field visits.

The instrument download schedule was updated each time data from an instrument were downloaded. The schedule accounted for the interval at which data were logged and the holding capacity of each instrument. The download schedule was maintained to ensure data were continually and efficiently managed.

3.4 Linked Relational Database

Over 4.6 million individual data measurements are housed and linked in the relational database for the period from 1999 through 2002. During the course of the 42-month study, data were collected from various automated equipment including:

- 28 thermistors (n ≅ 3.8 million data points)
- 12 networked rain gauges (n ≅ 1 million data points)
- 7 water elevation and velocity stations (n ≅ 400,000 data points)
- 2 air monitoring stations (n ≅ 170,000 data points)
- 1 solar station (n ≅ 370,000 data points)
- 1 wind station (n ≅ 43,000 data points)

Additional post-construction data were collected during summer 2003 and are noted in Chapter 6.0 and Appendix E.

Not all instruments were deployed for the entire 42 months; rather, as the project progressed, instrumentation was added, moved, or removed based on modeling and assessment needs. Instruments recorded information at five- or 10-minute intervals, producing a volume of data too large for spreadsheet manipulation; therefore, data were housed in a linked database structure. All data downloads were incorporated into a project database. Original field observations and laboratory QA/QC remarks were retained with the raw data. Field data were screened upon integration into the project database for obvious discrepancies prior to data reduction or analysis.

Individual relational databases were established for each instrument and linked together using a matching five-minute time interval. QC/QA measures, such as synchronizing the time intervals of instruments, formatting and matching software date–time formats, removing duplicate data points, removing corrupt data, tracking available data, and visually inspecting for integrity, were performed on all of the databases.

Due to the sheer number of data points, several software packages were used to manipulate, visualize, and analyze data. Microsoft Access was used to archive, organize, manipulate, and calculate simple summary statistics of the data, while MatLab then Systat were used for advanced statistics on the entire dataset. Microsoft Excel was used to analyze and plot separate portions of the data, such as individual events. Spatial data, including monitoring locations, land cover, hydrography, design drawings, topography and drainage networks were managed in separate geographic information system files (AutoCAD and ArcView).

3.5 Data Corrections and Substitutions

Unforeseen circumstances occasionally caused data loss (e.g., instrument malfunction, download malfunction, instrument theft or damage, lightning). When back-up instrumentation was not deployed in parallel, some substitution of data from nearby locations was necessary.

Pyranometer sensitivity declined during 2002. Alternate data was obtained from a matching unit maintained by Michigan State University at the Kellogg Biological Station (Hickory Corners, Michigan) approximately 40.2 km northeast of the project site. Comparable data from both units prior to 2002 confirmed satisfactory agreement between measurements and did not compromise the average hourly solar input data used for modeling.

3.6 Empirical Data Analysis

Data assessment focused on two key goals: 1) assessing the thermal balance of runoff passing through the stormwater treatment system components; and 2) comparing receiving stream thermal stability before and after construction of the stormwater treatment system related to the changes in the thermal load of runoff.

The combination of a high degree of pipe storage and limitations of the monitoring device at the sediment forebay inlet location made it necessary to make some assumptions about stormwater volume entering the forebay. Two distinct stages occurred following storms. In the first stage, direct measurement of the runoff entering the forebay was possible with the instrument. In the second stage, when stored water in the pipe upstream of the treatment system slowly drained into the forebay over a period of hours to days, the area/velocity meter was not sensitive enough to compute a flow. However, because the rate of elevation change in the forebay and wet pond were equal by design, it is implied that the more readily measured outflow rate of the wet pond (via a level logger) was controlling the drainage rate of the pipe. Consequently, the contribution of pipe-stored water in the second stage of discharge to the forebay could be estimated. Figure 3-3 illustrates the two distinct stages of forebay inflow.

3.7 Mechanistic Modeling

Li and James (2003) used an energy budgeting approach to model urban stormwater thermal enrichment. An earlier conceptual framework was expanded to include flow in all generalized elements of urban stormwater drainage systems including: 1) overland flow; 2) closed conduit or pipe flow; 3) open channel flow; and 4) ponds and wetlands.

To simulate the stormwater drainage system in this study, the watershed was discretized into numerous sub-watersheds for use in the hydraulic model, PCSWMM. Sub-watersheds are connected via sewers (closed conduit elements), channels (open-conduit elements), and ponds and wetlands. To compute water temperature in the network, temperature changes in every hydraulic element are considered separately.

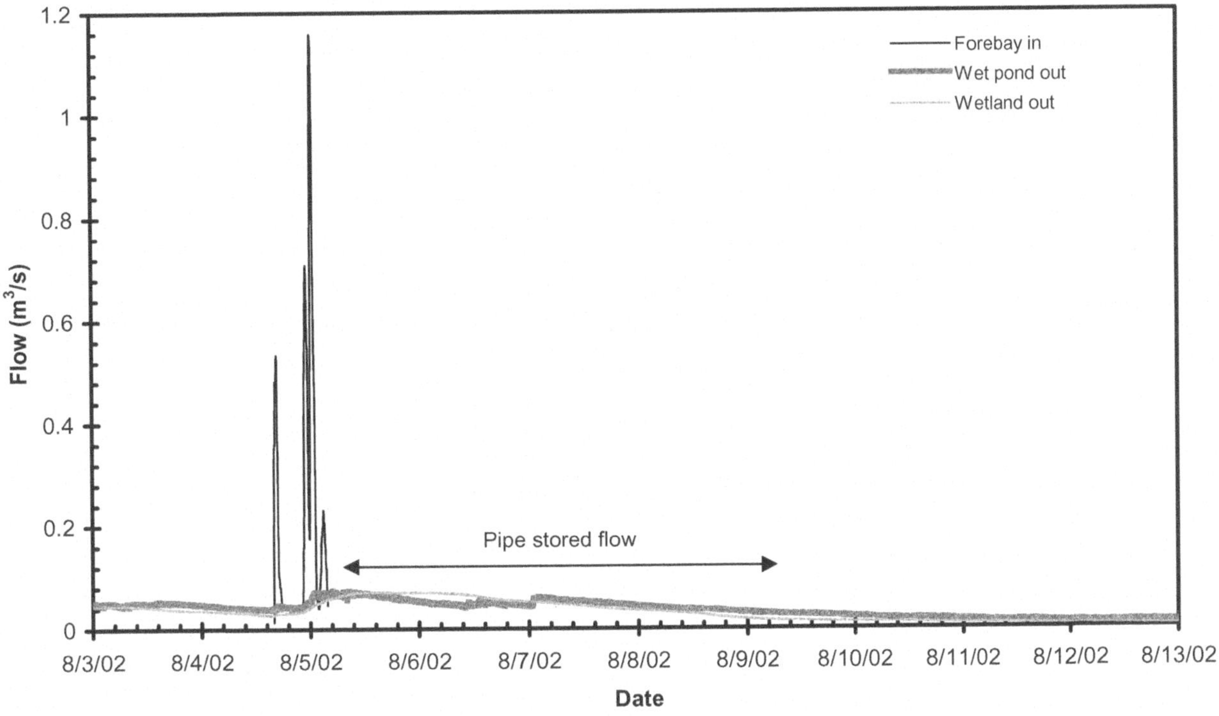

Figure 3-3: An Example of the Flow Profile of the Three Treatment Cells Responding to a Storm Event.

Modeled drainage areas were determined through the evaluation of 1999 satellite derived land use/land cover (LULC) data (Landsat 7), aerial photographs, topographic information, and digitized City of Portage storm sewer maps and housed in the GIS database. Percent impervious area was determined by converting standard LULC data.

3.8 Bioassessment

Benthic macroinvertebrates, periphyton, and fish were sampled to establish a baseline aquatic community assessment. The following sections describe the bioassessment methodology. The establishment of a cause and effect relationship between thermal regime change and community response was not attempted. Post-construction monitoring was limited to efforts conducted after one incomplete season of vegetation growth in the treatment system area. In addition, changes in hydrology and pollutant loading from the Consolidated Drain complicated any effort to isolate the impacts of individual stressors (e.g., thermal). Thus, Appendix C contains results and interpretation of these efforts.

3.8.1 General Methods

Bioassessments conducted for the project followed the Michigan's Qualitative Biological and Habitat Survey Protocols for Wadeable Streams and Rivers (Michigan Department of Environmental Quality, 2002) that are adapted from the U.S. EPA's *Rapid Bioassessment Protocols for Use in Streams and Wadeable Rivers* (1999). The bioassessment approach for addressing surface water quality issues associated with the Clean Water Act was first suggested by Karr (1981) in the development of the Index of Biotic Integrity. Since that time, bioassessments have been used in selected states, and several protocols have been developed by individual state and federal agencies. As opposed to the use of the more traditional water quality analyses for determination of the quality of a river system, bioassessments allow for the integration of chemical, physical, and biological factors into the ecosystem health analyses.

Sampling locations in this study consisted of six 100-m long reaches of stream (or drain) based on the requirements of the bioassessment protocol. Each reach was identified during the first field visit and was subsequently used for each sampling event thereafter. The six sampling locations included:

- Consolidated Drain
- Schuring Drain
- Portage Creek downstream of drain confluence
- Portage Creek at drain confluence
- Portage Creek upstream of drain confluence
- Portage Creek upstream of reference site

The upstream reference site was sampled in duplicate for QA/QC purposes. Details on the collection methods are presented as follows.

3.8.2 Benthic Macroinvertebrates

A composite sample was collected for each sampling location. A D-frame dip net with 500 μm mesh was used to sample 20 jabs within each 100-m reach. The jabs were collected from habitats (cobble, vegetation, snags, sand, etc.) in the proportion represented. The net was emptied into a white collection pan on shore where organisms were placed into a labeled plastic container using forceps or an eye-dropper, depending on organism size. The net was rinsed with deionized water when sampling locations changed. Benthic macroinvertebrate samples were collected in May and October of 2000 and 2001. A May 2002 sample was also collected. The samples were preserved with 95% ethanol and refrigerated. All samples were submitted to Phycotech, Inc. of St. Joseph, Michigan, for identification to the species level.

3.8.3 Periphyton

The same six bioassessment sites were sampled for periphyton. A composite sample was collected for each sampling location. Twenty samples per site were collected from substrate (rocks, submerged vegetation, submerged woody vegetation, sand, etc.) in the proportion represented within each reach. The samples were collected using a combination of eye dropper and toothbrush for scraping. Deionized water was used to rinse the sample into a labeled plastic container. Periphyton samples were collected in May, August, and November 2000. Samples were also collected in June, August, and October 2000. The spring 2001 sampling event was delayed until June, due to numerous wet weather events in May 2001 that did not allow for the monitoring protocol-specified recovery periods for the periphyton community prior to sampling. The samples were preserved with gluteraldehyde, wrapped in foil to inhibit light, and refrigerated.

The high-quality reference site identified for this study is a section of Portage Creek located upstream of the confluence with the Consolidated Drain. This site has had recent habitat improvements and represents a best-case condition for Portage Creek despite some recognized impacts from urbanization.

3.8.4 Fish

Fish collections were conducted on December 20, 2001, at the six locations sampled previously for benthic macroinvertebrates and periphyton. U.S. EPA rapid bioassessment protocols (RBPs) for stream fish sampling were used (1999). Fish were collected using a Smith-Root backpack electrofisher producing pulsed, direct current. Transects were sampled in an upstream direction for a maximum of 45 minutes or until the entire transect had been covered; block nets were not used given the relative inactivity of most fish in December. Fish were identified onsite and returned to the stream. One live individual was brought to the lab for identification and then returned to the stream.

Biotic community differences were investigated using metrics appropriate for this region's periphyton, macroinvertebrate, and fish communities. As with the other bioassessment procedures (benthic macroinvertebrates and periphyton), the intent of this portion of the study focused primarily on pre- and post-stormwater treatment facility construction conditions at *each site* rather than comparison of relationships between the sampling locations.

CHAPTER 4.0

TEMPERATURE EXCHANGE WITH CONTACT SURFACES AND WITHIN CONVEYANCES

This chapter presents observations and empirical evidence from the post-construction Consolidated Drain watershed that reveal basic principles of urban runoff thermal budgets. Information in this chapter establishes the basis for understanding the drivers of thermal enrichment in urban environments and the strategies necessary to mitigate thermal impacts of stormwater discharges to coldwater streams.

Note: Surface (pavement) temperatures as measured in this study do not indicate the capacity of heat storage in the subgrade materials. Heat exchange between surface and subgrade layers is beyond the scope of this chapter. Appendix D and Chapter 7.0 provide a conceptual basis and a modeling exercise that consider this heat exchange.

4.1 Thermal Enrichment Effects of Runoff

Characteristics of typical urban stormwater conveyance systems expose runoff to heat exchanges on impervious surfaces, in conveyance pipes, and in open conveyance channels. Generally, runoff is heated on impervious surfaces and in open channels exposed to direct solar inputs and is cooled in underground conveyance piping networks.

Runoff is thermally enriched when it absorbs heat stored in impervious surfaces (e.g., asphalt). Exposed impervious surfaces absorb heat during the day and release heat through the night. It is expected that daytime and nighttime runoff undergoes different degrees of enrichment depending on: 1) pre-event climatic conditions; 2) rainfall intensity; and 3) heat storage and exchange characteristics of impervious surface and underlying materials. Despite compounding factors for heat storage and exchange between runoff and impervious surfaces, evidence of thermal enrichment can be obtained simply by the observation that runoff from impervious surfaces is on average warmer than the rainfall temperature (assumed to equal air temperature) during the event.

As runoff moves through typical underground urban storm conveyance piping networks, the same principles of heat exchange with contact surfaces apply. Since these networks of pipes generally have a temperature cooler than air, the expectation is that runoff will undergo a net loss of heat in the underground conveyance. Therefore, despite similar complicated heat storage and exchange principles, basic evidence requires demonstration that runoff conveyed in pipes undergo a decrease in temperature.

Similar principles apply to runoff and baseflow in open conveyance channels that expose water to heat input due to direct solar exposure. Shaded channels provide the best protection from this input, but heat absorption also depends on water volume and channel shape. Demonstration of heat gained in open channels by runoff or baseflow requires evidence of increased temperature between separate points in the channel.

4.2 Methods

Surface temperature readings of paved surfaces were taken at the 30.8-ha Crossroads Mall (Portage, Michigan) during July and August 2002 using an infrared thermometer. Figure 4-1 shows the locations of the "NE," "SW," "Road," and "Drive" sample transects (represented as lines and dots around the mall perimeter).

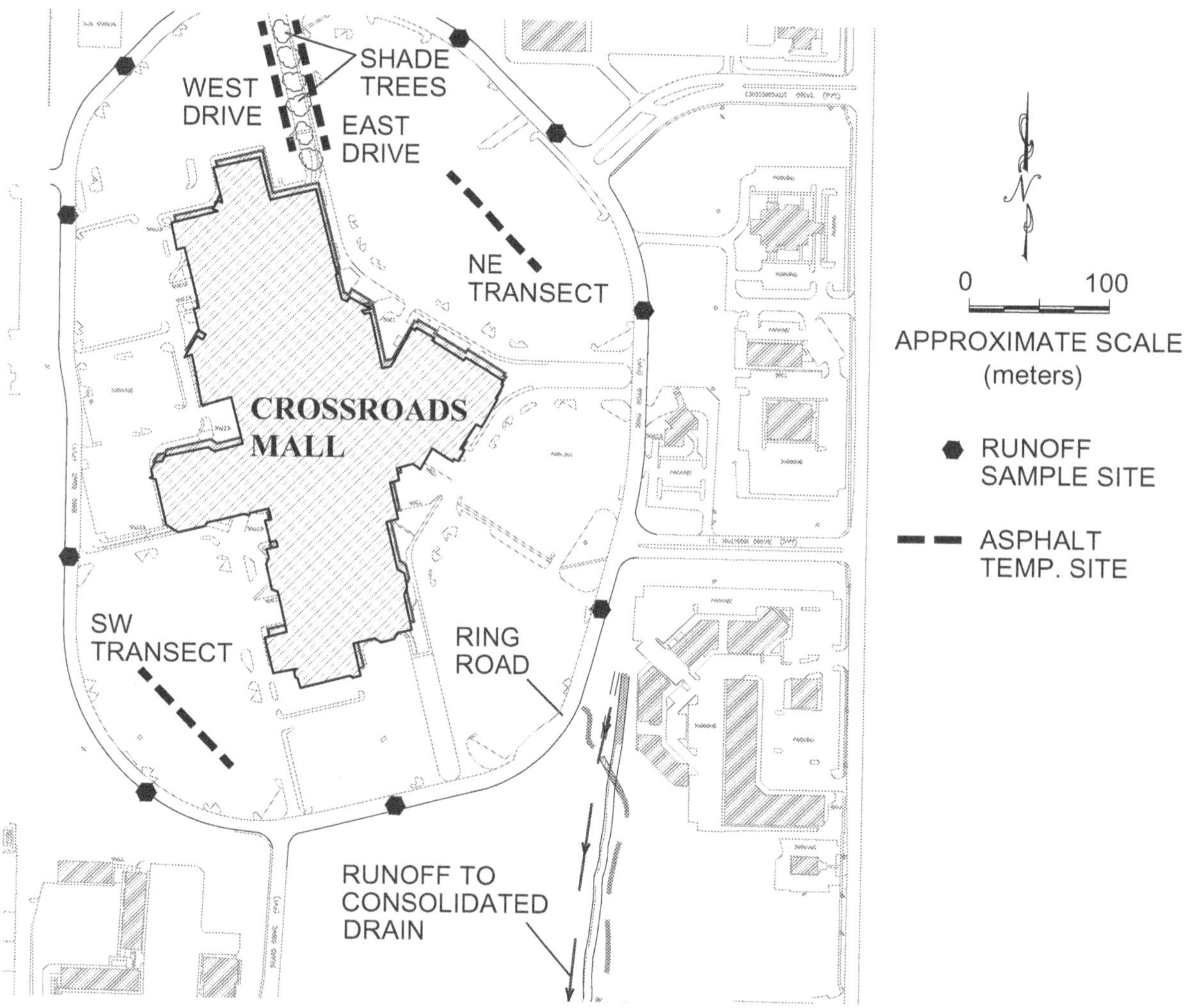

Figure 4-1. Pavement Surface Temperature Sampling Stations Located on the Grounds of the Crossroads Mall, Portage, Michigan.

Temperatures were measured on unshaded paved parking lot or road surfaces at the NE, SW, and Road locations. Transects NE and SW spanned fully exposed parking lots. The Road transect contained nine sample stations on the drive around the mall perimeter. The Drive station was located at the north entrance drive to the mall. Along this drive was a center island with a dense stand of mature 10-m tall shade trees. Pavement to the east and west sides of the island was exposed to direct sun at one time of day and was shaded at another. Transects were run on either side (east or west) of the drive. Temperatures were measured during conditions varying from morning to evening, clear to cloudy, dry to raining, and before, during, and after select rainfall events. Some temperatures were directly applicable to rainfall events and were coded as such (Table 4-1).

Table 4-1. General Observations Noted During Pavement Surface Temperature Monitoring at the Crossroads Mall, Portage, Michigan.

General Observation Codes (for Tables 4-2 and 4-3)	
A	After 20-minute light rain (no runoff generated)
B	After heavy rain (with runoff)
C	After 1-hour rain
D	Before 1-hour rain
E	Unusually cool day, sunny, with rain the previous night
H	During rain
Shadow	Temperatures taken in shadows of trees
Exposed	Temperatures taken on exposed pavement

Temperature of runoff was measured within the conveyance pipe that carries all mall catchment runoff to the Consolidated Drain (Figure 3-2). Runoff temperature at the discharge point to the Consolidated Drain was the product of initial rain temperature, energy gained (or lost) while in contact with impervious surfaces and the atmosphere, and energy lost (or gained) within the conveyance pipe.

Temperature of runoff from the Consolidated Drain watershed (including Crossroads Mall) was measured as it entered the sediment forebay. Characteristics of the treatment system (e.g., low elevation gradient) caused larger volume runoff events to back water up in the conveyance piping system. This condition resulted in pipe storage of water for a period of hours to days that slowly drained into the sediment forebay (see Section 3.6). Baseflow in the open Consolidated Drain channel before its confluence with the Schuring Drain originates as supplemental flow from a groundwater fed fountain (Figure 3-2). It is exposed to solar input in the wide and shallow low-flow channel.

4.3 Temperature Conditions

Characteristics of the contact surfaces and conveyance systems expose runoff to various heating and cooling forces. The following sections describe their effects on runoff temperature.

4.3.1 Partially Shaded Pavement

Pavement surface temperatures measured at the Drive transect reflect the variation on surfaces that are exposed to full sunlight for only a portion of the day. Average pavement surface temperatures in shaded areas were up to 17.6°C cooler than unshaded areas on sunny days (entries 13 and 14, Table 4-2). Pavement temperatures in shaded and unshaded areas were most similar: 1) during morning hours prior to prolonged solar exposure (entries 11 and 12); 2) during or after rain events (entries 7 and 8, 9 and 10, 23 and 24); and 3) on cloudy days (entries 19 and 20).

Table 4-2. Pavement Temperature Statistics for Drive Transect North of Crossroads Mall, Portage, Michigan.
The East and West Sides of Drive are Shaded by Trees for a Portion of the Day (see Table 4-1 for a description of codes).

Entry Number	Date Time	Average Temperature (°C)	SD (°C)	Min (°C)	Max (°C)	Count	Side of Drive	Conditions	Codes
1	7/16/02 12:25	28.9	0.7	28.1	29.6	5	west	sunny	shadow
2	7/16/02 12:30	46.0	0.4	45.4	46.5	6	east	sunny	exposed
3	7/16/02 15:50	40.9	1.2	39.9	42.7	5	west	sunny	exposed
4	7/16/02 15:53	33.6	1.7	31.5	36.2	5	east	sunny	shadow
5	7/23/02 10:30	23.9	0.2	23.5	24.1	6	west	sunny	shadow
6	7/23/02 10:33	36.3	0.5	35.7	37.0	6	east	sunny	exposed
7	7/29/02 13:21	29.4	1.3	28.4	32.1	6	west	cloudy	A
8	7/29/02 13:24	29.8	0.8	28.8	31.3	6	east	cloudy	A
9	7/29/02 15:28	26.6	0.2	26.3	26.8	6	west	cloudy	B
10	7/29/02 15:30	26.1	0.5	25.5	26.9	6	east	cloudy	B
11	7/30/02 9:07	29.8	0.6	29.1	30.4	6	east	sunny	exposed
12	7/30/02 9:12	24.2	0.4	23.7	24.8	6	west	sunny	shadow
13	7/30/02 17:07	48.4	0.8	47.1	49.2	6	west	sunny	exposed
14	7/30/02 17:10	30.8	1.1	29.3	32.2	6	east	sunny	shadow
15	8/12/02 11:42	29.1	0.5	28.1	29.4	6	west	sunny	shadow
16	8/12/02 11:45	41.0	1.4	39.2	42.6	6	east	sunny	exposed
17	8/12/02 16:06	34.6	2.5	31.6	37.2	6	east	sunny	shadow
18	8/12/02 16:08	46.5	1.3	45.2	48.3	6	west	sunny	exposed
19	8/13/02 15:42	34.4	1.3	32.4	35.7	6	west	cloudy	D
20	8/13/02 15:44	31.0	1.7	29.3	33.4	6	east	cloudy	D
21	8/13/02 18:08	33.4	1.3	31.5	35.0	6	west	sunny	C
22	8/13/02 18:11	25.4	0.3	25.1	25.9	6	east	sunny	C
23	8/22/02 6:58	22.3	0.1	22.1	22.5	6	east	cloudy	H
24	8/22/02 7:00	22.3	0.1	22.1	22.5	6	west	cloudy	H

4.3.2 Exposed Pavement

Readings from the three exposed sampling areas (NE, SW, and Road) were grouped for basic statistical analyses (Table 4-3). Generally, temperatures of fully exposed pavement were higher than pavement that was partially shaded during the day (e.g., compare Table 4-2 entry 3 with Table 4-3 entry 4). On sunny days, temperature of fully exposed pavement peaked as high as 52.8°C (entry 11, Table 4-3). Pavement temperature was lowest during morning and nighttime hours and highest during mid-afternoon hours (Figure 4-2).

Table 4-3. Pavement Temperature Statistics for Exposed Transects Including NE, SE, and Road at the Crossroads Mall, Portage, Michigan.

Entry Number	Date Time	Average Temperature (°C)	SD (°C)	Max (°C)	Min (°C)	Count	Run #	Codes
1	7/16/02 11:00	44.1	1.3	45.3	41.2	9	1	
2	7/16/02 12:15	45.9	0.8	46.7	44.9	5	2	
3	7/16/02 12:30	47.3	0.4	47.9	46.9	6	3	
4	7/16/02 15:50	49.6	0.9	51.3	48.0	19	4	
5	7/23/02 10:29	35.4	0.8	36.2	32.1	21	1	E
6	7/29/02 13:30	32.1	0.6	33.2	30.4	21	1	A
7	7/29/02 15:20	28.9	0.7	30.2	27.6	21	2	B
8	7/30/02 8:28	26.8	0.8	28.1	25.4	21	1	
9	7/30/02 16:35	50.2	0.9	51.7	48.7	21	2	
10	8/1/02 8:50	31.3	0.8	32.6	29.9	21	1	
11	8/1/02 14:20	52.8	1.2	55.9	50.4	21	2	
12	8/7/02 10:05	29.8	1.2	31.6	28.1	21	1	
13	8/7/02 15:18	45.0	2.8	47.5	33.9	21	2	
14	8/7/02 19:16	36.5	1.5	38.2	32.5	21	3	
15	8/12/02 11:48	43.3	0.8	45.1	42.4	21	1	
16	8/12/02 16:10	48.1	0.9	50.2	47.1	21	2	
17	8/13/02 15:15	42.2	1.8	45.1	38.9	15	1	D
18	8/13/02 18:00	31.7	1.3	34.1	29.1	21	2	C
19	8/21/02 8:32	31.3	0.1	31.4	31.1	6	1	
20	8/21/02 15:37	45.1	0.8	46.6	44.1	6	2	
21	8/21/02 20:35	30.2	0.6	31.4	29.3	21	3	
22	8/22/02 6:42	23.5	0.3	23.8	22.9	21	1	H

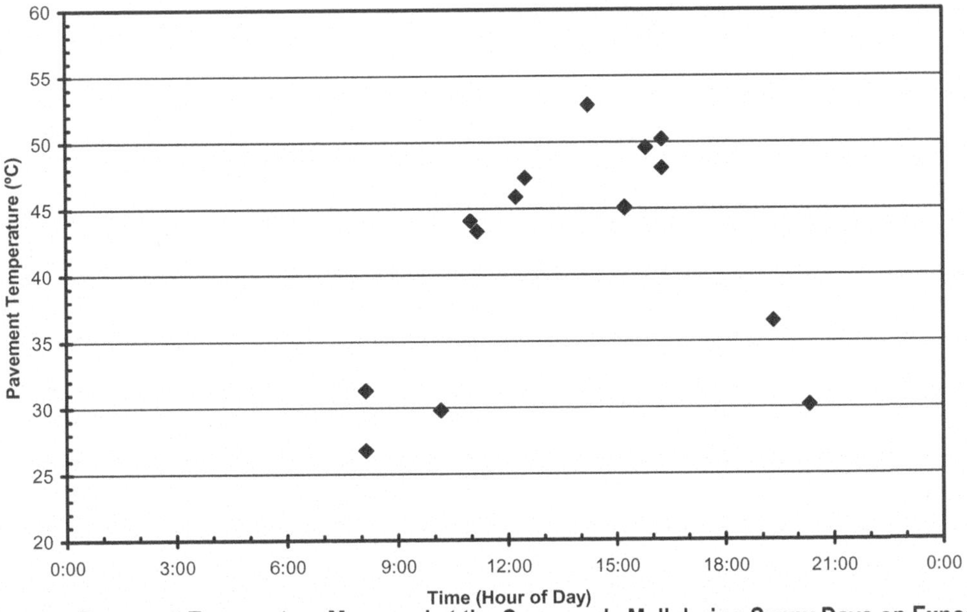

Figure 4-2. Instantaneous Pavement Temperature Measured at the Crossroads Mall during Sunny Days on Exposed Pavement Surfaces (based on Table 4-3).

4.3.3 Exposed Pavement vs. Air Temperature

Exposed pavement temperature was consistently higher than air temperature and reached its peak during periods of maximum solar intensity (Figure 4-3). Pavement temperatures at or near air temperature occurred during early morning hours, on cloudy days, and during or after rain.

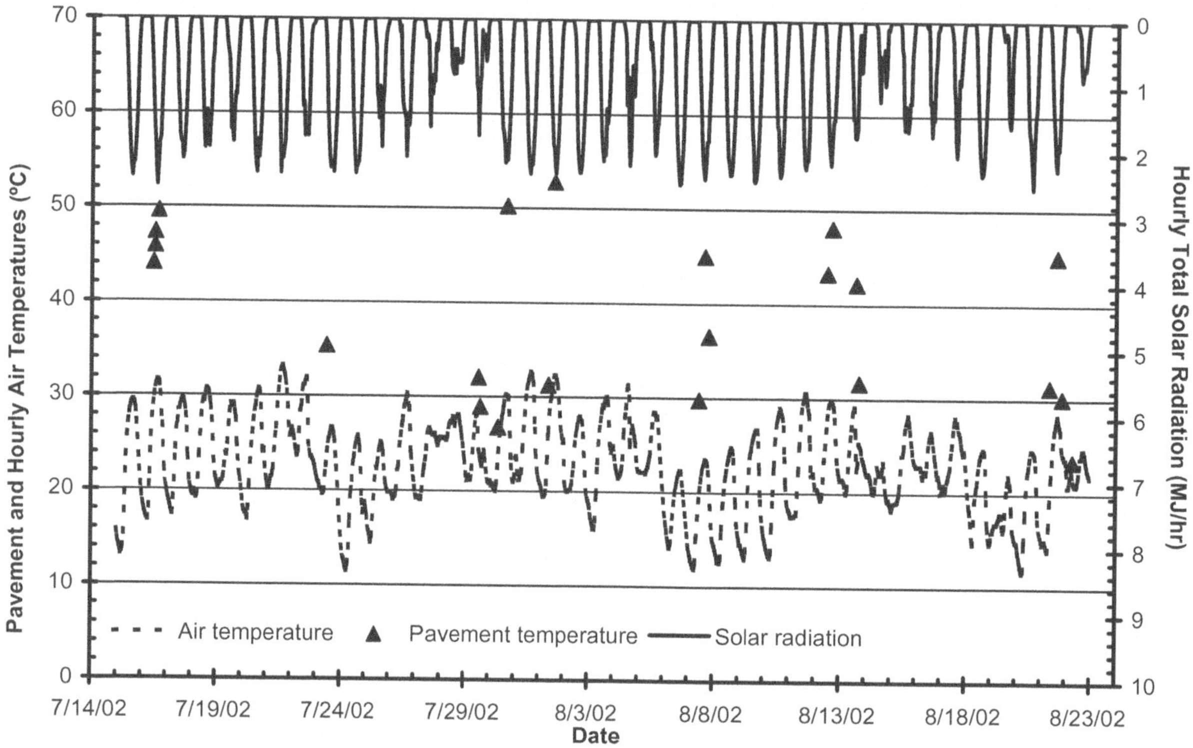

Figure 4-3. Instantaneous Pavement Surface Temperature Measurements, Hourly Air Temperature, and Hourly Total Solar Loading at Crossroads Mall, Portage, Michigan.

4.3.4 Runoff Temperature

Figure 4-4 illustrates the difference between peak (maximum) temperature of runoff and initial temperature of runoff (initial temperature assumed to be equal to air [Picksley and Deletic, 1999]). Runoff temperature was measured at the mall parking lot curb prior to entering the conveyance pipes leading to the drain. Twelve storm event runoff temperature differences were calculated for storms that occurred between August 2002 and September 2002. Temperature differences ranged from 0.5-4.3ºC during storms that occurred most often in the early morning (00:00 to 07:00 hours) and mid-afternoon (13:00-17:00 hours). Differences were always positive, indicating that the runoff temperature peak was always warmer than air temperature. In addition, review of individual events revealed that weather changes associated with rainstorms generally caused a rapid drop in air temperature with the onset of rain. However, while air temperature dropped, runoff temperature steadily increased, indicating that runoff was gaining thermal energy despite dropping air temperature (Figure 4-5). Maximum temperature differences also occurred during early morning and afternoon hours, indicating that runoff was being heated from stored energy (presumably in the pavement) during the day and night. Additionally, the greatest difference observed between the initial runoff temperature and the peak runoff temperature, particularly afternoon events, was only 4.3°C even though pavement surface temperatures regularly measured as much as 22°C higher than the air temperature.

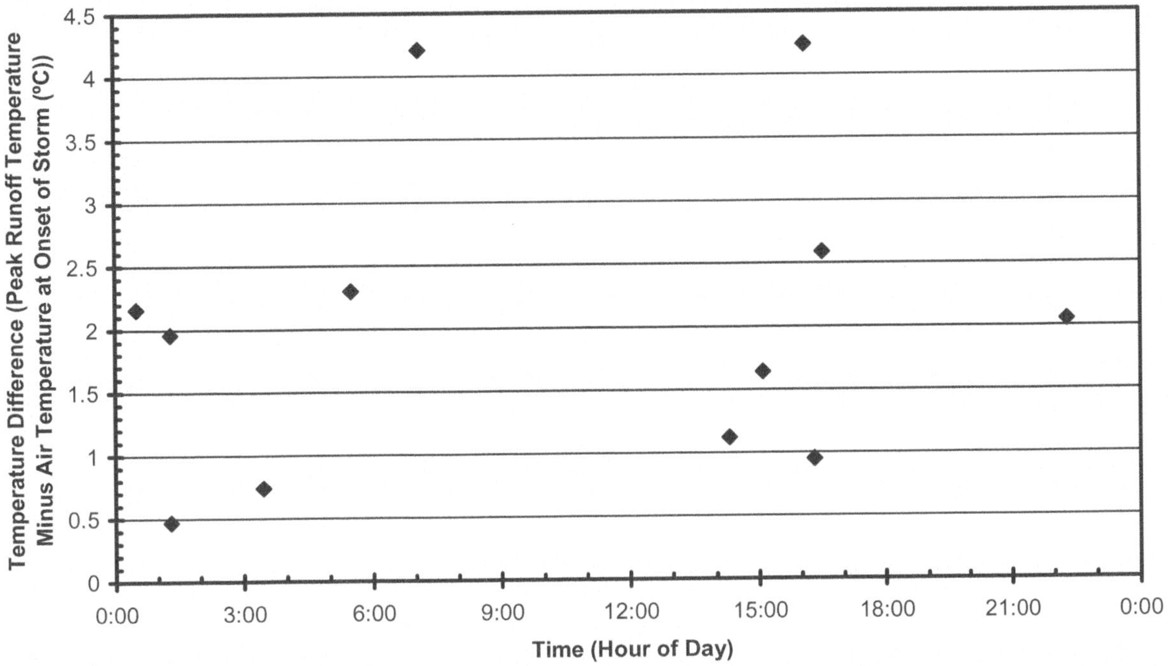

Figure 4-4. Temperature Difference Between Air (at the Onset of Runoff) and Peak Runoff Temperature During July and August 2002.

These findings demonstrate basic principles common to urban drainage networks including: 1) heat storage in impervious surfaces (in this case, mall parking lot pavement); and 2) runoff gains storing thermal energy during day and nighttime runoff events.

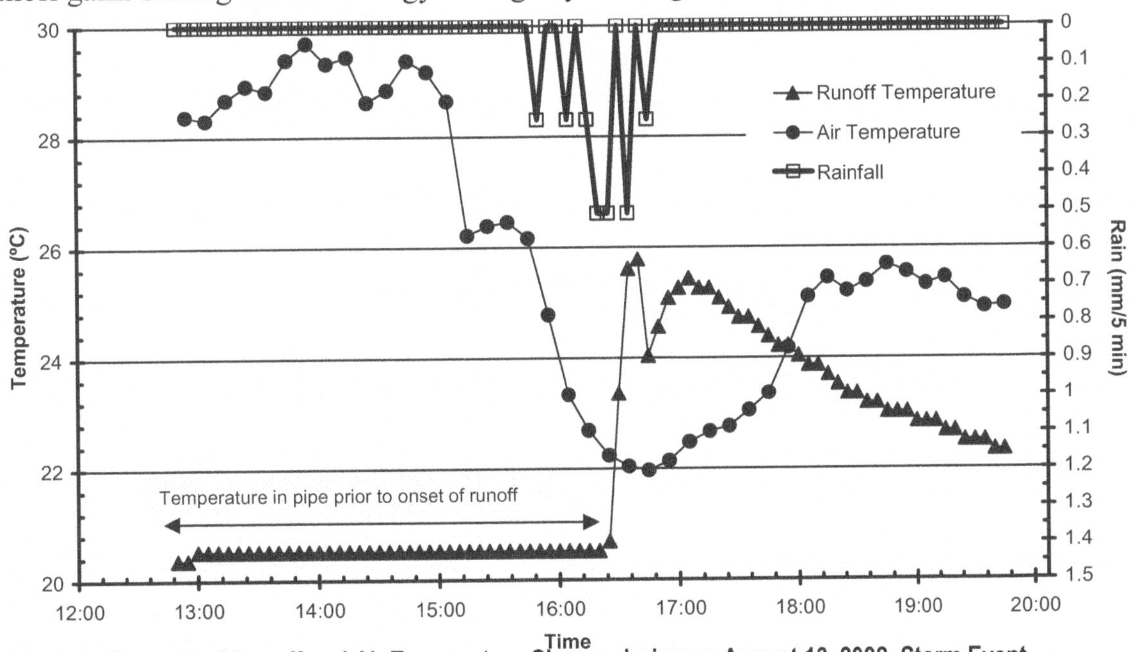

Figure 4-5. An Example of Runoff and Air Temperature Change during an August 13, 2002, Storm Event.

4.3.5 Temperature Change in Pipes

Following large rainfall/runoff events, characteristics of the treatment system caused a significant volume of water to back up in the upstream conveyance pipe network. Typically, runoff entered the sediment forebay at a rate as high as 1.19 m^3/s and then slowed to less than 0.028 m^3/s. Figure 4-6 shows that following high flow storm events, the runoff temperature gradually dropped until another storm occurred or until this pipe-stored flow ended. The changing temperature of the pipe-stored water represents the change in temperature of the runoff while stored.

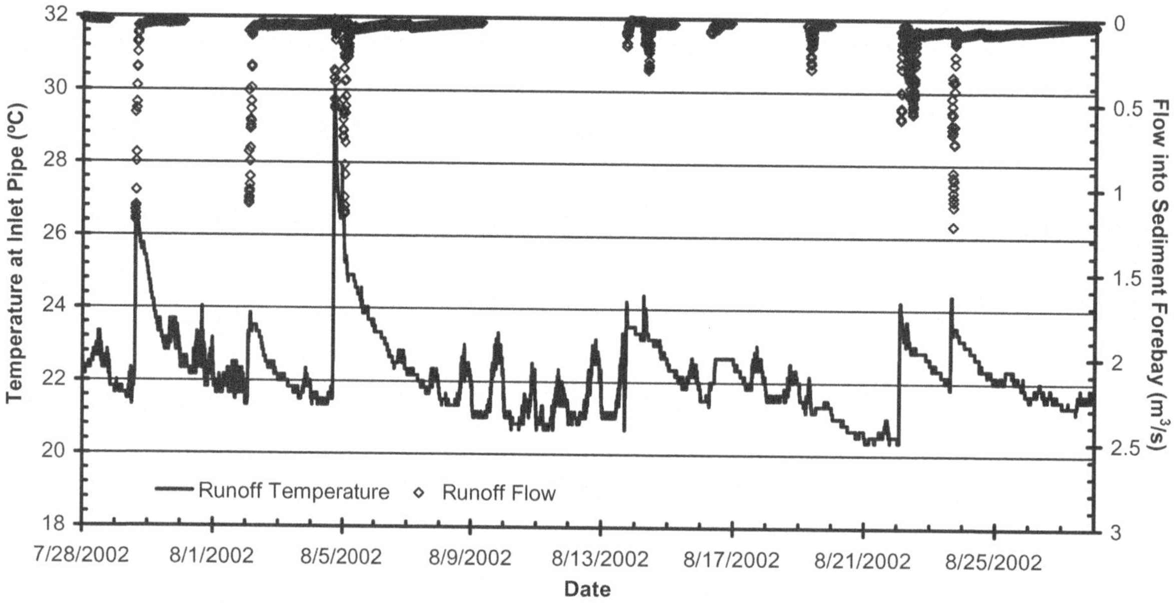

Figure 4-6. Runoff Flow and Temperature Entering the Sediment Forebay.

Post-event low flow of pipe-stored water measured as it entered the sediment forebay gradually dropped in temperature from the original runoff temperature peak (which was regularly at about 24°C and as high as 30°C). When pipe-stored flow ceased, the temperature became most heavily influenced by diurnal temperature fluctuations that normally occurred within the sediment forebay.

4.3.6 Open Channel Conveyance Temperatures

Baseflow from the groundwater-supplied fountain of the reflecting pond/ice rink entered the open Consolidated Drain floodway channel north of the system's treatment cells. Initial temperatures were near groundwater temperature but warmed as it passed through the channel before its confluence with the downstream Schuring Drain. Water entering the open Consolidated Drain channel at the reflecting pond was cooler than air temperature (Figure 4-7, rink – air). The diurnal temperature variation was high at the furthest downstream point in the channel prior to its confluence with the Schuring Drain and the Wetland outflow (Figure 4.7, drain – air). Water in the Consolidated Drain channel always had a higher temperature than air (as much as 13°C). The difference can be attributed to the high degree of exposure to direct sunlight in the channel during conveyance due to lack of riparian shading that had yet to develop one year after construction. Though the supplemental flow entered the channel near groundwater temperatures, it became warmer than air, indicating that solar heating is occurring as it passed through about 500 m of open channel.

The Schuring Drain baseflow was dominated by groundwater and its temperature difference from air is also shown in Figure 4-7. The Schuring baseflow did not heat in the same manner as the Consolidated Drain because it maintained a shorter, deeper, and more narrow channel; higher baseflow volume; and a higher degree of shading from the riparian vegetation (both on the adjacent channel banks as well as with the heavy growth of rooted aquatic macrophytes in the channel bed).

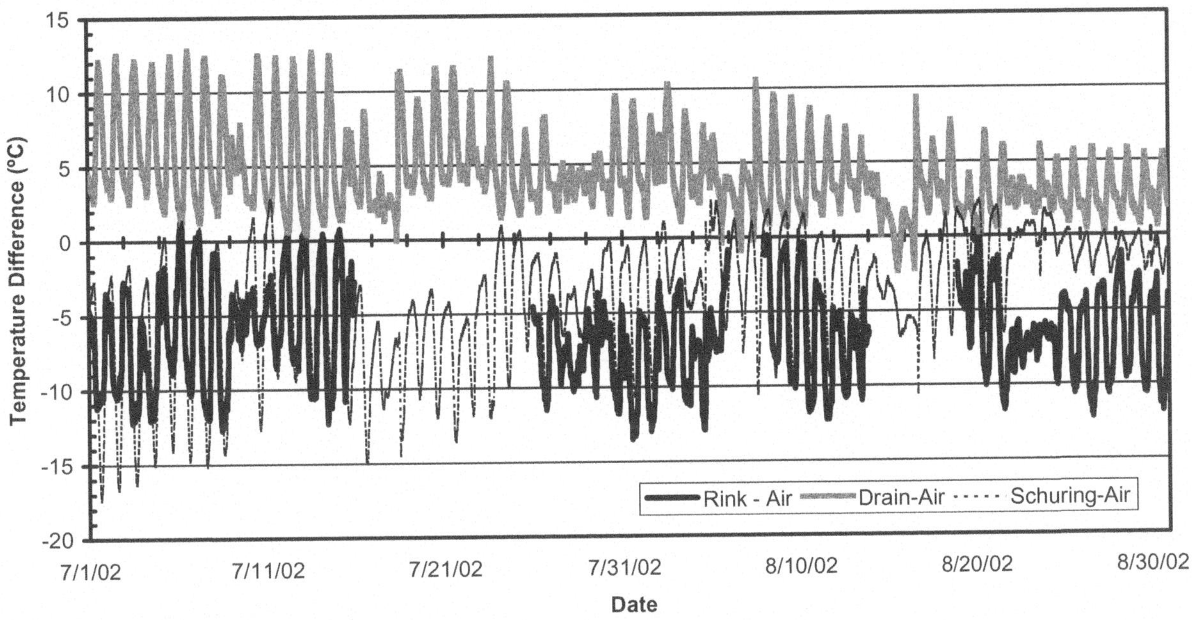

Figure 4-7. Temperature Differences Between Air and the Fountain/Rink Supplemental Baseflow, Consolidated Drain Open Channel Baseflow and Schuring Drain Open Channel Baseflow (10-minute Intervals).

Baseflow warming in the Consolidated Drain conveyance channel demonstrated that varying degrees of thermal enrichment can occur in urban stormwater conveyance systems. Both the Schuring Drain and Consolidated Drain baseflows are comprised of groundwater input, yet the Consolidated Drain underwent considerable diurnal warming due to a lack of riparian cover, which resulted in water temperatures well above air temperature. This is incompatible with the temperature regime of coldwater streams.

4.4 Support for Basic Tenants of Thermal Enrichment

Basic principles of thermal enrichment in urban runoff and associated conveyance systems were demonstrated by observations of the Consolidated Drain thermal regime. Evidence for the monitored period revealed that:

- Pavement surfaces absorb, store, and release heat, causing runoff thermal enrichment
- Runoff gains heat from pavement during the day and night
- Shaded pavement surfaces are cooler than exposed surfaces

- Pavement temperature is highest during mid-afternoon on sunny days
- Runoff temperature is warmer than air temperature
- Runoff loses heat in underground conveyance piping systems, but generally at low flows
- Groundwater baseflows in open conveyances gain heat due to exposure to direct solar input. This channel warming is considerably higher than in channels with substantial riparian and instream vegetative shading.

The Consolidated Drain watershed behaved in a manner consistent with the empirical understanding of urban runoff thermal enrichment. However, numerous spatial and temporal variables ultimately drive urban runoff heat exchange. Further empirical investigations are presented in Chapters 5.0 and 6.0 to quantify the thermal loading associated with treatment cells (ponds and wetlands) of the stormwater treatment system. Modeling efforts in Chapter 7.0 (and Appendix D) target the mechanistic understanding of these heat variables with the combination of empirical and thermal loading evidence used to calibrate a generalized urban stormwater thermal model.

CHAPTER 5.0

TREATMENT SYSTEM EFFECTS ON STORMWATER THERMAL LOADING

This chapter addresses the question of how the regional treatment system can affect the thermal loading of stormwater it receives in summer months. The treatment system was designed primarily to remove common stormwater pollutants, such as solids, nutrients, and metals. The design intentionally included some thermal remediation elements (e.g., standing trees in the natural wetland) and additional shade vegetation. This chapter approaches the question from three angles: 1) the water temperature regime of different treatment cells of the system; 2) overall stormwater thermal loading change in the treatment cells during the post-construction modeling period; and 3) stormwater thermal loading change from individual storm events. At the beginning of the chapter, an introductory section provides an overview of the sources and potential remedies of stormwater thermal enrichment, followed by a section introducing the "post-construction modeling" period and definitions of terms used in this chapter for the stormwater treatment system.

5.1 Heat as a Pollutant

Unlike common stormwater pollutants such as nutrients, metals and pesticides, thermal pollution is unique in that the "pollutant" is energy. Anything in the environment that can transfer energy in the form of heat to stormwater is a potential pollution source, including ambient air and direct solar radiation. In conventional detention ponds, while other pollutants are being removed from the water column via various physical and biogeochemical processes, ponded water is exposed to the air and direct solar radiation that can continuously add heat in summer months. Thus, while stormwater is effectively separated from sources of other pollutants by treatment facilities, it is still subject to the influences of two ubiquitous heat sources, the air and the sun. Barring mechanically cooling detained water, options to mitigate thermal loading and separate stormwater from these heat sources are limited. This is especially true for the Consolidated Drain stormwater treatment system in summer months and likely a similar constraint for other systems in northern temperate climates.

The Consolidated Drain stormwater treatment system consists of two conventional open water detention ponds and an enhanced natural wetland with dense vegetative cover. As such, it provides an opportunity to study the thermal behavior of stormwater runoff in treatment ponds. The analysis first characterizes the temperature regime of the three treatment cells during the "post-construction modeling" period and secondly, calculates thermal load and flow mass balances for 11 stormwater events monitored during this same period (July 29, 2002, to September 8, 2002). Data analysis illustrates how the system thermally responds to surface runoff from the Consolidated Drain watershed and how characteristics (e.g., shading and infiltration) of the system dictate the system's ability to mitigate or exacerbate stormwater thermal impacts.

5.2 Post-construction Modeling Period

Intensive monitoring of the treatment system during the summer of 2002 included a period of 42 days (July 29 through September 8) during which the last cell of the treatment system (the wetland) progressed from dry (having no outflow) to wet (continuously discharging treated stormwater) and back to dry (Figure 5-1).

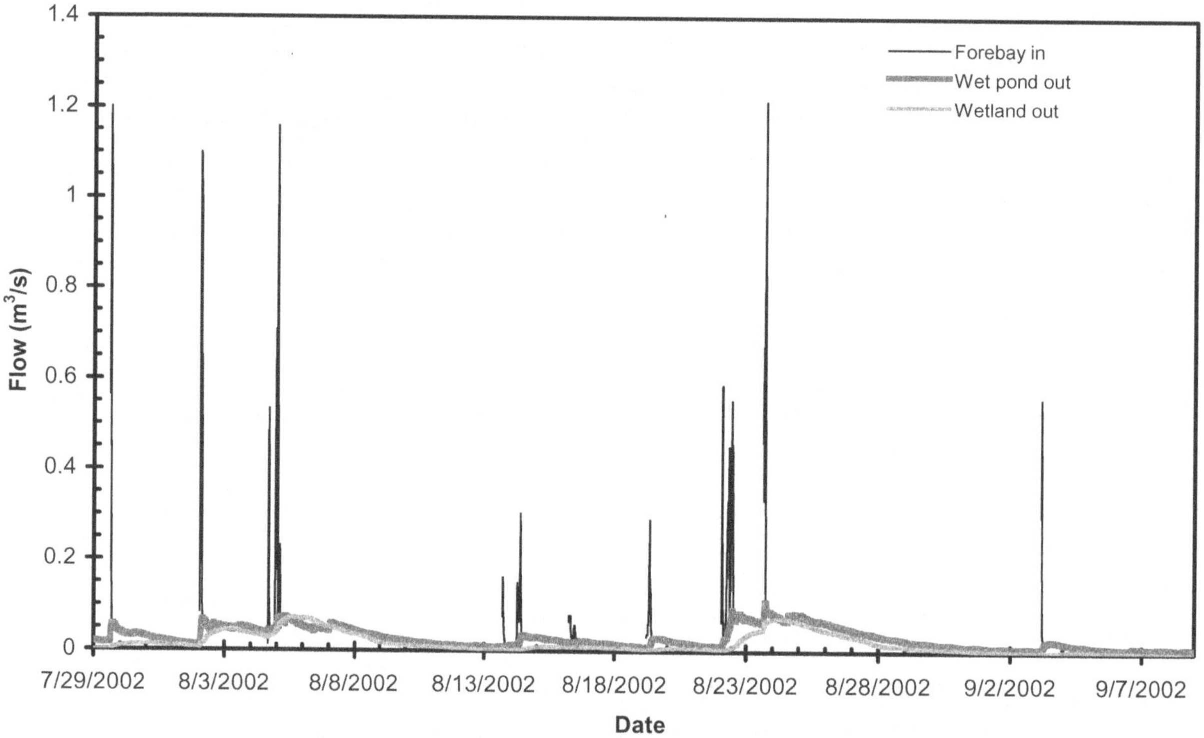

Figure 5-1. Flow Rates of the Treatment System During the Post-construction Modeling Period.

Because of the storage capacity provided by the treatment system, there is a time lag between the onset of a storm event and the exit of treated stormwater from the system via the third (wetland) treatment cell. This time lag causes the overlapping and mixing of runoff produced by successive storms in the system, complicating analyses aimed at examining the system response to individual storm events or a group of storm events taking place in a predetermined time interval. The 42-day dry-to-dry period during the summer of 2002, however, presented an opportunity to study the system's thermal treatment characteristics for the entire period and for individual storm events within the period. No outflow from the wetland signifies the completion of the treatment of all runoff collected by the system from previous storms and where the remnant storage would infiltrate or evaporate unless a new storm brought more runoff. Henceforth, this dry-to-dry period will be referred to as the post-construction modeling period.

Temperature and flow through the system were recorded at several locations including the:

- Stormwater pipe located on the bottom of the sediment forebay inlet (forebay in);
- Middle of the sediment forebay at a mid-water column depth (sediment forebay);
- Outflow point of the wet detention pond (wet pond out); and
- Outflow point of the wetland (wetland out).

Figure B-1 in Appendix B shows the locations of this instrumentation including the location of air temperature monitoring equipment in the wetland treatment cell.

During this post-construction modeling period, the average air temperature was 21.6°C with the maximum temperature reaching 32.9°C. There were 11 rain events that had total rainfall greater than 0.254 mm, four of which started in the afternoon hours (12:00 to 22:00 hours). The total precipitation during this period was 12.07 cm. The following analysis focuses on the thermal treatment effectiveness of the system and the mechanisms behind the thermal changes within the system that occurred during this period.

5.3 Temperature Regime of the Treatment System

Figure 5-2 shows a typical temperature profile of the three treatment cells responding to an afternoon storm event. Three stages of temperature change can be distinguished from the plot. The first stage is the period immediately prior to the onset of the storm event (15:50 hours, August 4, 2002). The steady, low flow in the stormwater pipe entering the forebay represents the dissipating pipe storage from the previous storm event and other unaccounted discharges to the pipe upstream of the Crossroads Mall in the Consolidated Drain watershed. This flow (forebay in) had the lowest temperature among all five locations. The summer afternoon air temperature (air 16:00, August 4, 2002) rose to 31.9°C, 75 minutes before the rain.

With a completely open water surface to absorb the summer heat from the sun and ambient air, the sediment forebay (forebay) had the highest temperature among all three treatment cells, reaching 27.4°C and still rising. Outflow from the wet pond (wet pond out) and the wetland (wetland out) had temperatures in between the stormwater pipe flow and the sediment forebay. The outflow from the largely open wet pond (with ponded water depths averaging 1.25 m at full storage capacity) had a lower temperature than the forebay (2.5 m average depth at full storage) but higher than the wetland outflow. This trend persisted through the post-construction modeling period under dry weather conditions. This suggests a cooling effect of the wet pond and the wetland on the water passing through them with the wetland being more effective than the wet pond.

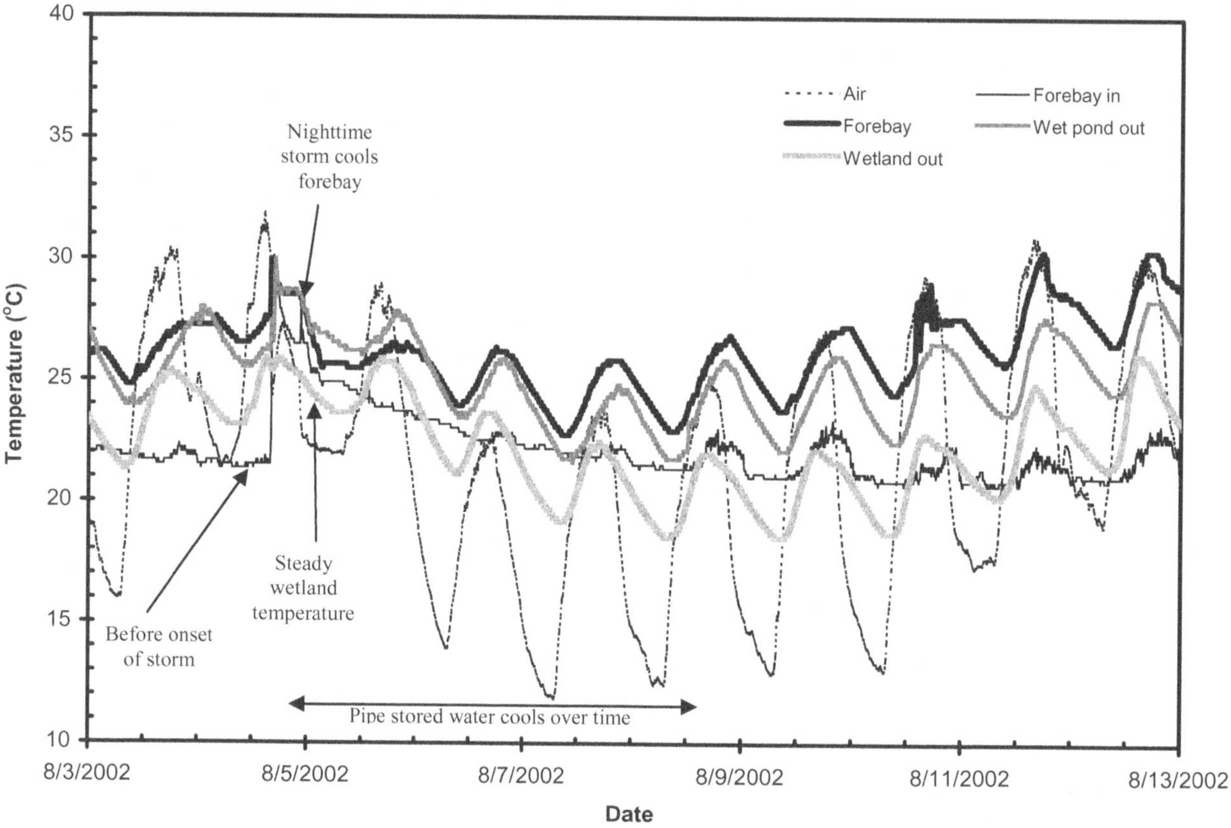

Figure 5-2. Example of Temperature (Five-minute Interval) Comparison Among the Three Stormwater Treatment Cells.

The second stage of temperature change in Figure 5-2 corresponds to the duration of the afternoon storm event (7.11 mm in 60 minutes beginning 15:50, August 4, 2002). The treatment system, especially the sediment forebay, responded quickly to the rainfall, reflecting the character of a highly impervious landscape on stormwater hydrology. There was an abrupt 8.5°C increase for the stormwater pipe inflow temperature, which can be attributed to the heat transfer from hot impervious surfaces in the watershed to the stormwater. Within an hour, the stormwater inflow raised the temperature of the sediment forebay and wet pond to the same level as the inflow. With its detention capacity, the wet pond held the storm runoff and the water temperature became steady for about four hours. It started to decline along with air temperature as the latter resumed its daily cycle after the storm. Wetland outflow held virtually unchanged throughout this stage and was the lowest until at about 22:30, when the air temperature dropped further toward its daily minimum. There was a second storm beginning late that night. It was a lingering drizzle. It increased the stormwater pipe inflow temperature briefly (likely due to the heat stored in pavement in the watershed since the cessation of the afternoon rain at 16:50) but soon started to bring the temperature down. This reflects the lower temperature of the rainwater at ambient air from the late night storm. This newly added runoff with a relatively low temperature and low flow rate was probably the cause of the divergence of the temperature in the forebay and wet pond (starting at about 01:00, August 5, 2002) when the water temperature of the sediment forebay dropped faster than that of the wet pond.

The third stage of temperature change in Figure 5-2 started when the late night storm stopped. This is a post-event period where the treatment system was exposed to additional thermal loading from the sun and the surrounding environment while holding the runoff water for an extended period. The stormwater pipe continued to drain stormwater runoff from storage. As seen in Figure 5-2, pipe storage flow draining into the forebay gradually decreased in temperature over a period of several days. Meanwhile, water in the sediment forebay, the wet pond, and the wetland resumed their daily temperature cycle paralleling daily air temperatures.

Two trends are identifiable in this third stage. The first is the rapid increase of water temperature in both the sediment forebay and the wet pond. The second is the increasing wetland outflow temperature over time compared to the inflow of the sediment forebay. The first trend suggests the heating of open waters of the sediment forebay and wet pond. The second trend suggests that while the wetland always provides for the cooling of water flowing through it from the wet pond, over time during the summer months wetland outflow temperature eventually rises to be higher than the minor inflow received by the forebay from the diminishing pipe storage. Monitoring data indicate that when this happens, the wetland outflow is usually at its lowest levels. However, the maximum outflow temperatures are still lower than the air temperature (Figure 5-2).

Figure 5-3 illustrates the differences between the daily temperature highs and lows of the system components during the same event period shown in Figure 5-2.

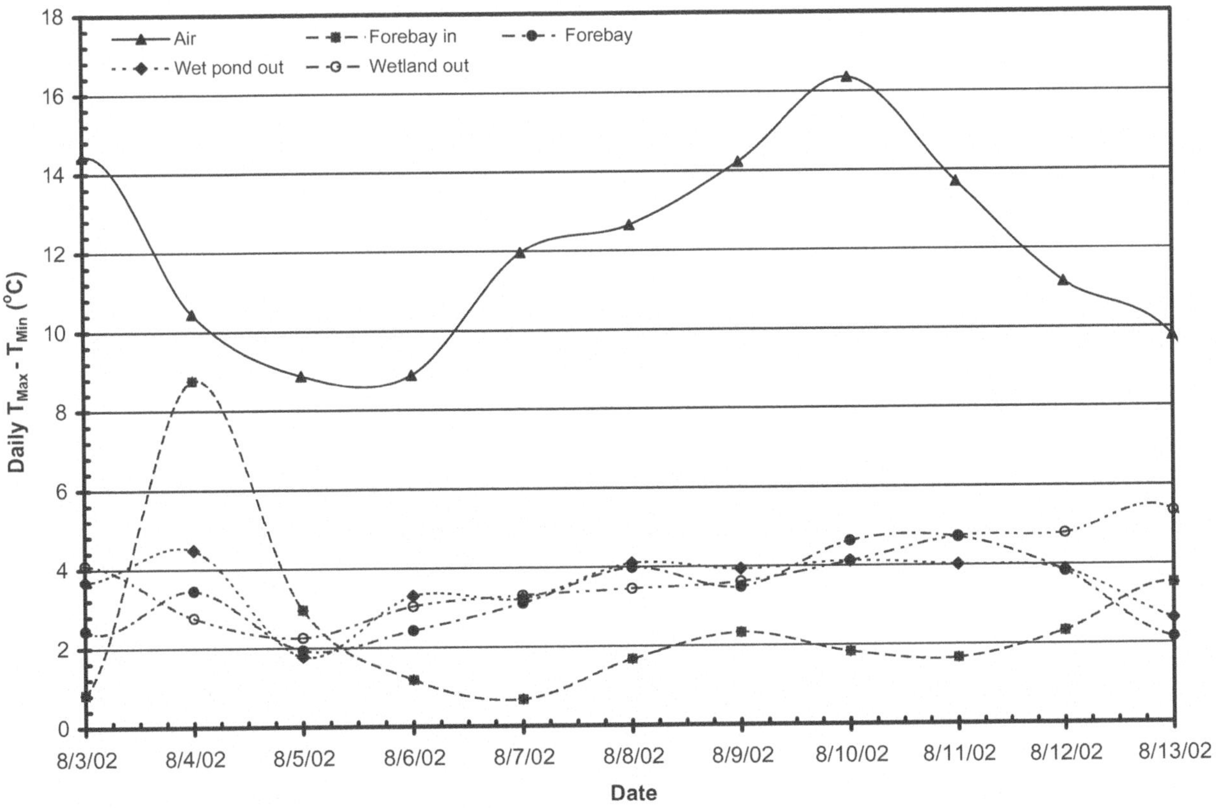

Figure 5-3. Daily Temperature Fluctuations (Daily Maximum – Daily Minimum) of the Three Stormwater Treatment Cells.

Figure 5-3 demonstrates the daily temperature fluctuations of the three treatment cells. The three-stage temperature change pattern is also observable. Due to the high heat capacity of water, daily temperature fluctuations of the three cells are much lower than that of the air. Nevertheless, on the event day (August 4, 2002) forebay inflow displayed the highest daily fluctuation (8.8°C), indicating stormwater thermal enrichment in the watershed. After the cessation of the rain events, inflow to the forebay fluctuated the least due to the underground piping system that collects and temporarily stores part of the runoff. Water in the three treatment cells exhibited similar daily fluctuations (within 1°C of each other until the next storm event on August 13, 2003). Over the entire period, daily temperature fluctuation for water detained in the treatment cells increased during the storm (except the wetland) due to the warm inflow of storm runoff. This was followed by an decrease immediately after the storm, likely due to the relatively stable air temperature (Figure 5-2). As the air began to resume its high daily fluctuation, water temperature fluctuations in the treatment cells increased slightly until the next storm event.

For the entire post-construction modeling period, temperature changes of the three treatment cells are best illustrated by comparing the average daily temperature records of each cell with the ambient air temperature (Figure 5-4). Subtracting the daily average temperature of air from the cells, it becomes clear that stormwater detained in the

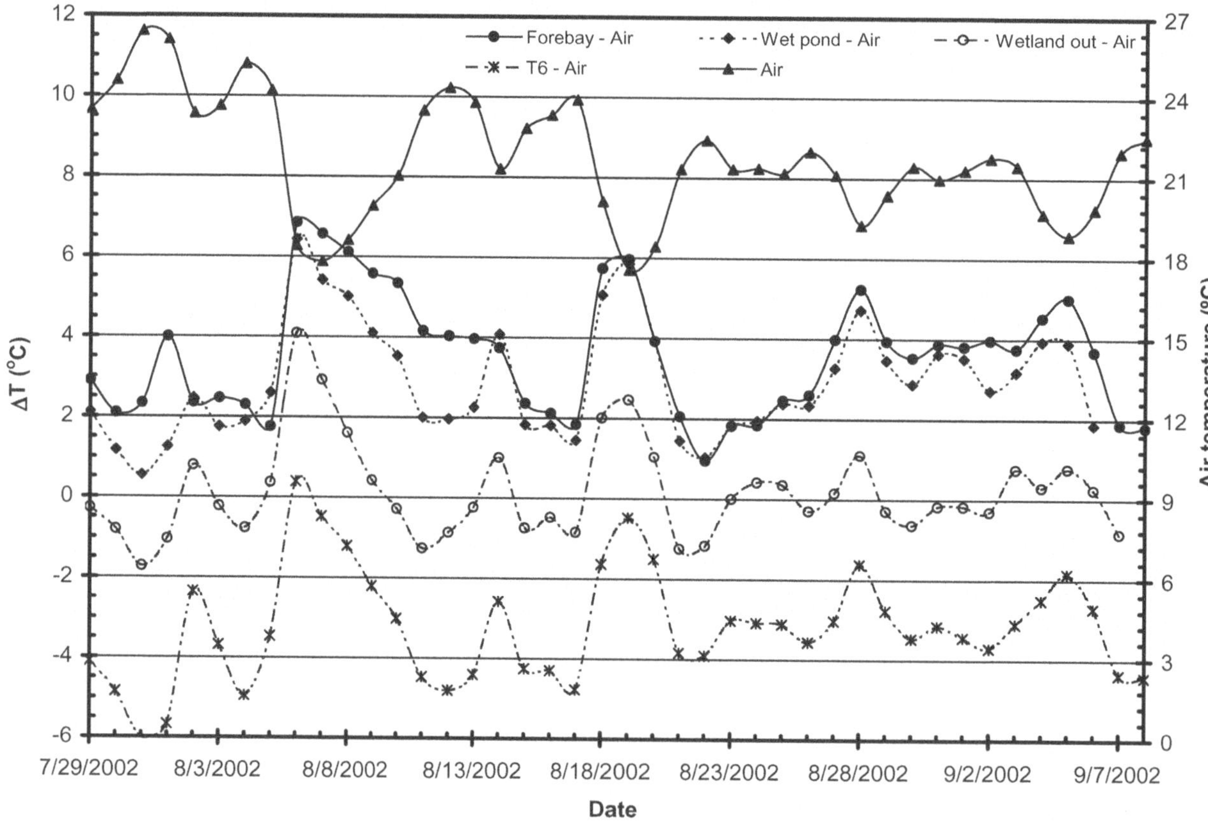

Figure 5-4. Average Daily Temperature Difference Between the Air and the Treatment System Cells (and Thermistor T6 in Portage Creek Upstream of its Confluence with the Consolidated Drain) During the Post-construction Modeling Period.

sediment forebay and the wet pond was always warmer than the ambient air (average ΔT forebay: 3.6 °C, range: 1.0-6.8 °C; wet pond: 2.9°C, range 0.5-6.4°C). This confirms the heating effect or additional thermal loading to the collected stormwater runoff in the two ponds during the detention period due to solar inputs. In contrast, the wetland was able to mitigate this thermal load and maintain an outflow temperature close to ambient air on a daily average basis (ΔT average: 0.1°C, range: -2.8 °C to 4.1°C).

The most obvious and significant differences among the three cells that can be used to explain this temperature regime difference demonstrated by Figure 5-4 are vegetative cover and water depth. A well-shaded water surface prevents direct solar radiation from reaching the standing water and enables it to equilibrate with the ambient air temperature. Deeper ponds have smaller surface area to volume ratios, rendering these ponds less efficient in heat dissipation. The sediment forebay is essentially an open pond without any meaningful vegetative cover. The wet pond has a few tall cottonwood trees and a variety of wetland scrub/shrub vegetation that provides limited shading. Ponded areas in the wet pond, intentionally designed to provide habitat diversity, are basically devoid of any significant vegetative cover. In contrast, a layer of densely populated grasses, shrubs, and patches of trees cover the wetland cell. These provide substantial shading in the wetland, allowing average daily water temperatures to approach those of ambient air. In addition, the two ponds maintain a water depth three to four times greater than the wetland. The wetland also has a smaller overall surface area to its volume. In essence, the system configuration is such that the two open ponds absorb heat more efficiently than the wetland during the day and release heat more slowly during the night, resulting in higher water temperatures. The close reverse correlation between the air temperature and the temperature differences between the treatment cells and the air is another indication of the heat absorbing behavior of the stormwater detained by the treatment cells.

Figure 5-4 also illustrates the temperature difference between the air and thermistor 6 (T6 – Air), which is located in Portage Creek about 1000 m upstream of the confluence with the Consolidated Drain channel. Portage Creek almost always has a lower daily temperature than ambient air (ΔT average: -3.2°C, range: -6.0°C to 0.4°C), signifying the features of a coldwater aquatic habitat. The implication is that although with the help of the wetland, the treatment system was able to bring the outflow daily average temperature close to that of ambient air, the outflow would still have had a warming effect on the receiving stream had it drained directly into the creek. Instead, wetland outflow drains to the Consolidated Drain channel where there is groundwater baseflow mitigating this potential warming effect. Chapter 6 of this report further examines this in the context of the thermal impact of the entire Consolidated Drain watershed on Portage Creek.

5.4 Effects of the Treatment System on Stormwater Thermal Load

This section focuses on the change of stormwater thermal load in the Consolidated Drain treatment system during the post-construction modeling period in the summer of 2002. Two questions are addressed: 1) overall, was the treatment system effective in reducing the thermal load in stormwater; and 2) what were the primary mechanisms that accounted for the thermal load change in the system? Section 5.4.1 introduces the general mechanisms and processes that can affect stormwater thermal loading, and Section 5.4.2 provides a detailed analysis of the thermal loading calculation results for the post-construction modeling period.

5.4.1 Heat and Flow Components of the System

The temperature of any pond system is the result of heat exchange between the water and the surrounding environment (e.g., Thomann and Mueller, 1987; Bedient and Huber, 1992).

Principle components of this heat exchange include heat gains through short-wave radiation, long-wave radiation, and heat conduction (sensible heat) from the atmosphere to water. Heat losses include long-wave radiation emitted by water, evaporation, and heat conduction from water to the atmosphere. For an exposed stormwater detention pond, however, a significant form of heat exchange is the flow of water (including direct rain deposit) into and out of the system. There are primarily two reasons for this. First, water has a high heat capacity. Any change in water temperature causes an increase or decrease in heat load to the system. Second, the nature of an urban stormwater treatment system determines that there will be a substantial increase of water volume in the system whenever a sizable storm event occurs. This increase of water volume represents a significant change of the total heat load (and hence the temperature regime) of the system, particularly the sediment forebay and the wet detention pond of the City of Portage treatment facility.

On the other hand, flow components of the system are relatively few, involving only inflow and outflow, direct rain deposit, evaporation, infiltration, and groundwater input (if the water table is sufficiently high). Groundwater elevations were examined prior to stormwater treatment system construction (K&A and URS, 1999). Piezometer readings indicated that no significant groundwater input to the system would occur when the system is discharging water at the wetland outlet. Therefore, only inflow and outflow, direct rain deposit, evaporation, and infiltration were considered in water balance calculations.

5.4.2 Computation of Thermal Loads

The mass balance Equation 5.4.2-1 was used to reach a balanced water volume budget for the forebay–wet pond combination, the wetland, and the treatment system as a whole, for the post-construction modeling period (July 29, 2002, to September 8, 2002).

$$(\textit{Inflow Volume} + \textit{Rain Deposit}) - (\textit{Outflow Volume} + \textit{Evaporation} + \textit{Infiltration}) = 0 \qquad (5.4.2\text{-}1)$$

Calculated water volumes for each of the three components are presented in Table 5-1. Negative numbers indicate water losses. In Equation 5.4.2-1, inflow, rain deposit (direct rainfall on the system), and outflow were calculated based on measured parameters only. Evaporation, including evapotranspiration, was calculated using monitored meteorological data from the system (solar radiation, wind velocity, air temperature, and humidity) and the Penman equation coupled with a Meyer estimate of lake evaporation (Penman, 1948; Meyer, 1944; Bedient and Huber, 1992). Various studies (e.g., Kadlec and Knight, 1996) have suggested that with vegetative cover, wetlands have less evaporation due to reduced solar radiation as a result of vegetative shading while transpiring more through the vegetation. In general, these two effects offset each other and the total evapotranspiration of wetlands is similar to that of lakes. There is no readily available, widely accepted estimation method for calculating wetland evapotranspiration based on meteorological data. Therefore, the evaporation rate calculated based on lake evaporation is used for all three treatment and cell combinations in Table 5-1. Infiltration was calculated to account for the residual volume to balance Equation 5.4.2-1.

Uncertainty in the volume calculations arose mostly from the estimation of evaporation with the Penman-Meyer approach. Different estimation methods yield somewhat different results for evaporation and hence, infiltration. For example, using equations recommended by the

American Society of Civil Engineers (Jensen et al., 1990) would result in evaporation estimates about 20% higher than the Penman-Meyer approach for the post-construction modeling period. Consequently, the infiltration volume calculated with Equation 5.4.2-1 for the entire treatment system would decrease by 11% and the total water volume reduction accounted for by infiltration would decrease from 82% to 73%. These numbers show that uncertainty in evaporation estimation does not substantially change water budgeting. The Penman-Meyer approach was chosen for this study because it yielded a small negative infiltration value (loss of water into the ground; Table 5-1) for the forebay–wet pond combination, consistent with the water level in the forebay–wet pond always being higher than the area groundwater table.

Table 5-1. Volume (m^3) Losses (-) and Gains (+) in the Treatment System During the Post-construction Modeling Period.

Contribution (m^3)	Forebay–Wet Pond	Wetland	System Total
Inflow	111,895	105,671	111,895
Outflow	-105,671	-64,652	-64,652
Total Loss	-6,223	-41,019	-47,243
Rain Deposit	6,884	5,664	12,548
Evaporation [a]	-11,609	-9,551	-21,160
Infiltration [b]	-1,498	-37,132	-38,631

[a] Including transpiration by vegetation, based on Penman (1948) and Meyer (1944).

[b] Residual of inflow, outflow, rain deposit, and evaporation (See Equation 5.4.2-1).

Infiltration rates were derived from the residual volume reduction to further verify the calculations for infiltration volume. These calculated infiltration rates were compared to hydraulic conductivity test results for soil samples taken from the system prior to construction (K&A and URS, 1999). The forebay–wet pond combination had a calculated infiltration rate of 7.44×10^{-7} cm/s while the value for the wetland was 2.24×10^{-5} cm/s. Overall, the system had an area-weighted average infiltration rate of 1.05×10^{-5} cm/s. Soil samples from the system had hydraulic conductivities ranging from 1.9×10^{-7} cm/s to 2.8×10^{-4} cm/s. Calculated infiltration rates were well within this range. Furthermore, lithology investigations showed a confining layer of marl underneath most of the topsoil of the treatment system. This marl layer is most prominent in the western portion of the system (coinciding with the sediment forebay and the wet pond) and becomes thinner as it approaches the eastern side of the system (where the wetland is located). Directly below the marl layer is a sand unit. Therefore, the derived infiltration rates are considered reasonable estimates for the ponds and the system as a whole.

Thermal loads were calculated here for the post-construction modeling period using Equation 5.4.2-2, which is widely applied in surface water quality modeling and heat budget calculations (e.g., Thomann and Mueller, 1987; Bedient and Huber, 1992; Van Buren et al., 2000a):

$$W = Q \cdot \rho \cdot T \cdot C \cdot t \qquad (5.4.2\text{-}2)$$

where: W = thermal load (cal),
Q = flow rate (m^3/s),
ρ: water density (assumed to be constant at 1×10^3 kg/m^3),
T: water temperature (°C),
C: water heat capacity (assumed to be constant at 1 cal/g/°C),
t: time (s).

Equation 5.4.2-2 demonstrates that there are two ways for any stormwater treatment system to reduce the thermal load of runoff it receives—lowering water temperature (*T*) or reducing water volume (Q). Equation 5.4.2-2 further indicates that these two approaches are equally effective as volume has the same weight as temperature on the result of the equation. Consequently, it is possible for a treatment system to have a net thermal load reduction but still not achieve temperature mitigation (a higher outflow temperature than inflow), when water volume reduction is the primary mechanism of thermal load reduction. This is important because it implies that to achieve complete thermal mitigation, it may be necessary to deal with thermal loading, of which temperature is a factor, and temperature. This issue is further explored in later sections of this report.

Based on monitoring data and Equation 5.4.2-2, during the post-construction modeling period, there was a total thermal load of 2.585×10^9 kcal to the treatment system (i.e., entering the sediment forebay) through the stormwater pipe collecting runoff from the Consolidated Drain watershed (Table 5-2). This load was delivered by a total volume of more than 1.1×10^5 m^3 of stormwater to the sediment forebay. The conventional stormwater treatment cells (i.e., the sediment forebay and the wet pond) produced a minimal heat reduction of 1.3%. The wetland, however, provided a 43.9% reduction of heat over the output it received from the wet pond. This resulted in a wetland outlet temperature much lower than the forebay and the wet pond.

Table 5-2. Heat and Water Volume Change Through the Treatment System During the Post-construction Modeling Period.

System Location	Heat (× 10^9 kcal)	Volume (m^3)
Forebay In	2.585	111,895
Wet Pond Out	2.551 [a]	105,671
First Reduction [b] (%)	0.034 [c] (1.3)	6,223 (5.6)
Wetland Out	1.431	64,652
Second Reduction [b] (%)	1.120 (43.9)	41,019 (3.38)
Total Reduction [b] (%)	**1.154 (44.6)**	**47,243 (42.2)**

[a] Temperature data not available from 15:10, September 6 through 10:20, September 8.
[b] First reduction: from forebay in to wet pond out; second reduction: from wet pond out to wetland out; total reduction: from forebay in to wetland out (the entire treatment system).
[c] Actual values will be lower due to temperature data gap.

Two mechanisms of water volume and heat reduction can be called upon to explain the higher thermal treatment efficiency of the wetland. The first is a higher infiltration rate for the wetland and the second is the shading provided by the dense vegetation growing in the wetland.

Table 5-2 shows that the wetland accounted for most of the 42.2% total flow reduction by the system during the post-construction modeling period. A high infiltration rate enhances flow reduction and hence, heat load reduction (Equation 5.4.2-2). Shading directly reduces solar (short wave) radiation received by the system, leading to lower temperature extremes of water leaving the wetland compared to the air temperature (see Figure 5-4).

5.5 Individual Storm Events

A closer examination of the effects of the treatment system on stormwater runoff involves the evaluation of individual storm events taking place in the post-construction modeling period. As the focus on thermal enrichment is on the treatment cell scale in this assessment, a plug-flow approach was used. To achieve water volume balance for each storm, the following steps and assumptions were made:

- As before, it was assumed that inflow and outflow volumes of the system were storm runoff from the pipe feeding the sediment forebay, direct rain deposits on the treatment cells, evaporation, and infiltration.
- Calculated infiltration rates of 7.44×10^{-7} cm/s for the sediment forebay and the wet pond and 2.24×10^{-5} cm/s for the wetland were assumed to be constant throughout the post-construction modeling period. This is a reasonable approximation considering the underlying soil properties and the constant presence of water in these treatment cells during this period.
- Calculated daily evaporative losses were disaggregated into hourly values using the WDMUtil program from U.S. EPA's BASINS model (Version 3.0, U.S. EPA, 2001). Hourly values were then evenly divided into minutes.
- A storm runoff event was defined for the wet pond and the wetland with the beginning of the event marked at the time there was an abrupt increase of the flow on the corresponding hydrograph or immediately after the ending of the previous event. The end point of an event for each cell was identified at the point of time that all the inflow volume to the cell had exited the outlet of the cell, minus the evaporation and infiltration plus the direct rain deposit from the storm.

These assumptions resulted in discrepancies between the sums of the 11 individual event values and the overall values (Tables 5-1 through 5-3) for this period at only 8.0% and 8.6% in stormwater heat and water volume inputs to the system, respectively. For system outputs, the differences were at 1.5% and 1.7%, respectively. These differences were considered negligible.

As derived from Table 5-3, loss of water volume by evaporation, including evapotranspiration, on average accounted for 19.9% of the total inflowing volume while infiltration represented 40.8%. Both of these loss mechanisms were important factors in the water balance of the system. Infiltration, however, was more dominant.

Table 5-3. Heat and Water Volume Reduction by the Overall Stormwater Treatment System and Individual Components for the 11 Individual Rain Events During the Post-construction Modeling Period. Negative numbers (-) indicate heat or volume gained.

	Overall System Data								*Component Data*			
Event No.	**Heat In ($\times 10^8$) (kcal)**	**Heat Out ($\times 10^8$) (kcal)**	**Heat Reduction (%)**	**Inflow Volume (m^3)**	**Outflow Volume (m^3)**	**Volume Reduction (%)**	**Evaporation (%)[a]**	**Infiltration (%)[b]**	**Forebay-Wet Pond Heat Reduction (%)**	**Wetland Heat Reduction (%)**	**Forebay-Wet Pond Volume Reduction (%)**	**Wetland Volume Reduction (%)**
1	2.20	0.89	59.3	8,757	3,721	57.5	24.6	41.9	3.7	57.7	-8.61	53.5
2	3.70	3.31	10.4	16,384	13,672	16.6	12.5	20.1	-17.2	23.6	0.88	17.3
3	0.67	0.58	13.6	2,369	2,586	-9.1	4.7	17.4	-3.7	16.7	13.62	3.9
4	5.69	2.76	51.6	24,056	12,984	46.0	19.7	34.6	5.1	49.0	-7.82	41.4
5	0.25	0.09	62.7	1,065	420	60.6	43.7	41.7	6.3	60.2	-16.08	53.1
6	0.90	0.39	56.7	3,873	1,700	56.1	20.7	52.5	-2.1	57.5	-3.63	54.5
7	0.44	0.15	65.3	1,960	678	65.4	24.0	54.9	4.1	63.8	-3.67	64.1
8	0.64	0.22	65.4	2,971	1,134	61.8	24.9	57.4	-1.6	66.0	-3.51	60.4
9	3.65	3.06	16.2	15,927	14,261	10.5	4.7	19.1	-6.9	21.6	5.56	15.2
10	5.09	2.53	50.3	22,379	11,874	46.9	16.7	38.2	-1.3	50.9	-3.89	44.8
11	0.58	0.11	81.3	2,517	554	78.0	22.9	70.9	3.4	80.7	-8.36	76.0
Sum	*23.8*	*14.1*	--	*102,258*	*63,583*	--	--	--	--	--	--	--
Average	*2.16*	*1.28*	*48.4*	*9,296*	*5,780*	*44.6*	*19.9*	*40.8*	*-0.9*	*49.8*	*3.23*	*44.0*

[a] Evaporation as a percentage of the total inflow.

[b] Infiltration as a percentage of the total inflow.

While evaporation rates were similar across the three treatment cells, the wetland had a much higher infiltration rate than the forebay–wet pond combination. Consequently, most of the infiltration occurred in the wetland. Table 5-3 shows that the forebay–wet pond combination had a mere 3.23% volume reduction, compared to the wetland's 44.0%. Due to the exposure of detained water in the forebay and wet pond to summer solar radiation, these two cells combined had small or even negative heat reduction (i.e., heat gain) for the 11 events, with an average value of -0.9%, despite a positive average water volume reduction. In contrast, the wetland had a higher average heat reduction (49.8%) than the average volume reduction (44.0%), indicating that volume reduction was not the only mechanism that reduced heat in the wetland. A likely explanation is that the shading provided by the dense vegetation during the study period (late July to early September) in the wetland prevented detained stormwater from adsorbing heat from solar radiation. This allowed the evapotranspiration in the wetland to have a net heat reduction effect.

For the entire treatment system, there is a strong relationship between water volume reduction and heat reduction. Figure 5-5 shows that for the Consolidated Drain stormwater treatment system, heat reduction outpaced volume reduction during the post-construction modeling period, likely due to the shading and evapotranspiration in the wetland. This relationship demonstrates the importance of water volume reduction within the wetland as noted in Table 5-3. A net heat reduction could be achieved by the system as long as the water volume increase (due to direct rain deposit onto the treatment cells) did not exceed 11.6% of the inflow volume to the sediment forebay.

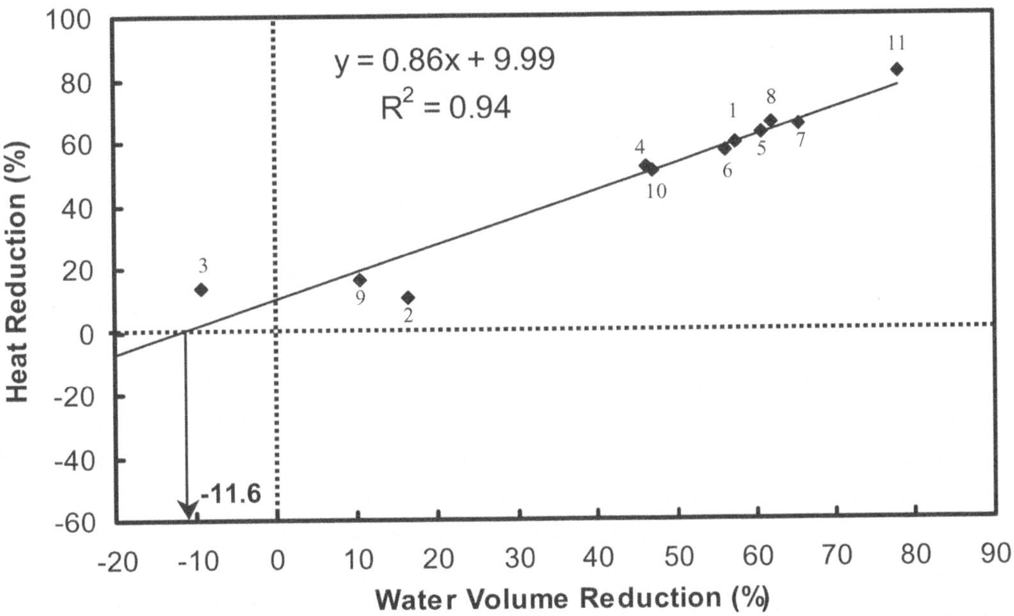

Figure 5-5. Relationship Between System Water Volume Reduction and Heat Reduction for the 11 Individual Events During the Post-construction Modeling Period for the Stormwater Treatment System. Numbers near the data points are event numbers, as shown in Table 5-3.

CHAPTER 6.0

STORMWATER THERMAL IMPACTS ON PORTAGE CREEK

This chapter examines the impacts of stormwater on Portage Creek, before and after the Consolidated Drain stormwater treatment system was constructed. Compared to Chapter 5.0, it focuses more on the thermal behavior of the entire Consolidated Drain watershed and the thermal regime changes in the receiving stream induced by stormwater. It also puts the investigation of the treatment system's thermal treatment efficiency into the perspective of watershed conditions of the pre- and post-construction periods. Examined here are:

- The effects of development on the thermal regime of coldwater streams (Section 6.1).
- Specific comparisons of summer thermal loading to Portage Creek from the Consolidated Drain watershed in all four years of monitoring (Section 6.2).
- Introduction of a temperature index, or temperature equivalent, to identify thermal loading contributing areas in a watershed (Section 6.3).
- Temperature regime changes in Portage Creek as a result of the construction of the treatment system and accompanying hydrological modifications in the watershed (Section 6.4).

6.1 Thermal Impacts from Watershed Development

In pre-development conditions, there are few areas that store heat in watersheds of coldwater streams. Surface runoff is substantially reduced through infiltration and vegetated canopy interception. Michigan streams are fed primarily by groundwater, which lowers the stream temperature during summer months (Wehrly et al., 1998). In addition, shading provided by riparian vegetation minimizes additional heating of the stream from direct solar radiation. In contrast, pavement and other constructed surfaces in the post-development watershed effectively absorb solar radiation. These transfer heat to surface runoff during rainfall events. Heated runoff received by the coldwater stream increases in-stream temperatures, potentially exerting thermal stresses on aquatic biota.

Traditional stormwater best BMPs, such as detention ponds, are designed to mitigate the eroding forces of stormwater runoff volumes, and to remove pollutants carried by runoff via settling or plant uptake. However, as shown in Chapter 5.0, detention ponds expose collected stormwater to additional thermal loading from ambient air and solar radiation, significantly increasing the temperature of detained stormwater. In the Consolidated Drain stormwater treatment system, a well-vegetated wetland as the final treatment step results in a system outflow that has, on average, the same temperature as ambient air.

To determine whether this meets design goals to mitigate post-construction thermal impacts on Portage Creek, this chapter focuses on a receiving stream evaluation from two perspectives:

- Thermal loading to Portage Creek from the entire Consolidated Drain watershed, (including the outflow from the treatment system, the Schuring Drain flow, and the flow from the Consolidated Drain channel upstream of its confluence with the Schuring Drain); and
- Temperature regime change in Portage Creek due to flow from the Consolidated Drain watershed.

6.2 Thermal Loading to Portage Creek (Pre-construction vs. Post-construction)

Summer thermal loading to Portage Creek from the Consolidated Drain watershed in all four years of monitoring is examined in this section for periods before and after construction of the treatment system. To define the relevant aspects of the multi-component Consolidated Drain watershed and associated analyses, a system of specific terminology is provided in Sections 6.2.1 and 6.2.2. These include a detailed description of the components of the watershed and the two-month comparative period, respectively. Subsection 6.2.3 presents the results of these analyses.

6.2.1 The Consolidated Drain Watershed

During the five-year (1999 through 2003) monitoring period of the Consolidated Drain watershed, the stormwater treatment system was constructed beginning in August 2000 and came online in mid-July, 2001. The first storm event that the system captured and treated took place on July 21, 2001. To examine stormwater thermal impacts on Portage Creek both with and without the treatment system, and to discern the effect of the treatment system on thermal loading to the creek, it is necessary to compare the thermal behavior of the Consolidated Drain watershed before and after the treatment system was constructed. Data from 1999 and 2000 are considered the pre-construction period while data after July 21, 2001, through 2003 are from the post-construction period.

The construction and operation of the treatment system changed the regional hydrology substantially. In pre-construction years of 1999 and 2000, the only area contributing a substantial amount of untreated stormwater to the Portage Creek was the Consolidated Drain watershed. This is equivalent to the post-construction treatment system's drainage plus the area now occupied by the treatment cells (Figure 3-2). Flow from the Schuring Drain in pre-construction years was directed into a private pond with only occasional overflows to the Consolidated Drain channel. Therefore, the pre-construction Schuring Drain contribution was assumed to be zero. As a result, in 1999 and 2000 the contribution of stormwater to the creek was assumed to be exclusively from the Consolidated Drain open channel with only the 187-ha watershed drained by the channel—the pre-construction Consolidated Drain watershed.

With construction of the treatment system, the Schuring Drain was routed away from the private pond and connected directly to the Consolidated Drain channel just upstream of the outlet of the treatment system (i.e., the wetland outflow; Figure 3-2). As a result, Schuring Drain flows, outflow from the treatment system, and treatment system overflows discharge to the Consolidated Drain channel and then to Portage Creek. The combination of the treatment system drainage and the Schuring Drain watershed became what the post-construction Consolidated Drain watershed. Overflow from the weirs near the Crossroads Mall into the Consolidated Drain channel during large storm events (mall overflow) were monitored.

6.2.2 The Two-month Comparative Period

A two-month period was chosen to be August 1 through October 1, a window of 62 days, in each of the four monitoring years (1999 through 2002) to examine the total thermal loading to Portage Creek from the pre-construction and post-construction Consolidated Drain watershed. The reason for choosing only the months of August and September for 2001 is that August 2001 was the first full month that the treatment system came online and September can still be considered a summer month of thermal loading concern (Galli, 1990). For 2002, detailed flow and temperature monitoring for various flow components in the post-construction Consolidated Drain watershed did not start until the beginning of August for reasons explained in Chapter 3.0. For 1999, August 2 was the start date for the monitoring work of this study. For 2000, the period of August 1 through October 1 was chosen for consistency for the cross-multi-year comparison. (A one-day adjustment was made [August 2 through October 2] in 1999 due to the absence of monitoring data on August 1.) Because this time period has essentially two full months of data over the four years, it is termed the August-September comparative period.

In 2002, the August-September comparative period overlapped with the dry-to-dry post-construction modeling period (July 29 through September 8) as discussed in Chapter 5. Both periods contained the entire month of August 2002. With the exception of three days toward the end of July, the entire dry-to-dry post-construction modeling period was included in the comparative time period. In addition, only three storm events that had hourly rainfall greater than 0.5 mm took place between September 9, 2002, and October 1, 2002. Wet pond and wetland outflow hydrographs also indicated that the surface runoff generated by these storms were very small compared to those occurring during the "dry-to-dry" period. Therefore, results about stormwater thermal loading from these study periods could be compared.

6.2.3 Pre-construction vs. Post-construction Thermal Loading

Additional flow monitoring equipment was provided through the City of Portage such that starting August 1, 2002, an intensive and detailed monitoring program (flow, temperature, and various pollutants) was initiated for the key hydrological components of the treatment system. Results from this monitoring program allowed for a precise breakdown of the thermal loading contributions from these components, namely the two detention ponds, the wetland, runoff from the Crossroads Mall, runoff from the treatment system watershed upstream of the mall, and runoff and baseflow from the Schuring Drain and the Consolidated Drain channels.

In 1999, 2000, and 2001, flow and temperature were measured in the Consolidated Drain open channel, just upstream from its confluence with Portage Creek (Figure 3-2). During 2001, a flow instrument malfunction made it necessary to assume average baseflow conditions in the Consolidated Drain channel between August 2, 2001, and August 10, 2001. There was no rainfall during this period. Therefore, no thermal loading from stormwater runoff was missed as a result of this necessary assumption. Although the treatment system was operational in August and September 2001, its individual components were not monitored as they were in 2002. Therefore, a precise breakdown of component contributions is not directly possible for 2001. The treatment system and contributing area water volumes and associated heat loads for the comparative pre- and post-construction period are provided in Tables 6-1 and 6-2, respectively.

Table 6-1. Runoff Volume (m³) to Portage Creek During the Two-month Comparative Period of August-September for the Pre- and Post-construction Consolidated Drain Watersheds.

				Contributing Drainage Areas					
Treatment Period	**Year**	**Data Period**	**Rainfall (mm)**	**Treatment System**		**Mall Overflow ($\times 10^5$ m^3)**	**Schuring Drain ($\times 10^5$ m^3)**	**Total Consolidated Drain ($\times 10^5$ m^3)**	**Unknown Component ($\times 10^5$ m^3)[a]**
				Forebay Inflow ($\times 10^5$ m^3)	**Wetland Outflow ($\times 10^5$ m^3)**				
Pre-construction	1999	8/2-10/2	109	-	-	-	0[b]	1.76	-
	2000	8/1-10/1	160	-	-	-	0[b]	3.62	-
Post-construction	2001	8/1-10/1	277	Not measured	Not measured	Not measured	Not measured	5.10	-
	2002[c]	8/1-10/1	140	1.230	0.688	0.034	1.85	3.14	0.569

[a] Unknown component refers to unaccounted load contributions to the Consolidated Drain channel calculated by total volume minus measured volumes from contributing drainage areas (Schuring, mall overflow, and wetland outflow).
[b] Assumed (see Section 6.2.1).
[c] Year 2002 includes component volumes not measured in previous years.

Table 6-2. Runoff Thermal Load (kcal) to Portage Creek during the Two-month Comparative Period of August-September for the Pre- and Post-construction Consolidated Drain Watersheds.

				Contributing Drainage Areas					
Treatment Period	**Year**	**Data Period**	**Rainfall (mm)**	**Treatment System**		**Mall Overflow ($\times 10^9$ kcal)**	**Schuring Drain ($\times 10^9$ kcal)**	**Total Consolidated Drain ($\times 10^9$ kcal)**	**Unknown Component ($\times 10^9$ kcal)**
				Forebay Inflow ($\times 10^9$ kcal)	**Wetland Outflow ($\times 10^9$ kcal)**				
Pre	1999	8/2-10/2	109	-	-	-	0[a]	2.71	-
	2000	8/1-10/1	160	-	-	-	0[a]	6.06	-
Post	2001	8/1-10/1	277	Not measured	Not measured	Not measured	Not measured	8.99	-
	2002[b]	8/1-10/1	140	2.79	1.48	0.08	3.00	5.47	0.92

[a] Unknown component refers to unaccounted load contributions to the Consolidated Drain channel calculated by total volume minus measured volumes from contributing drainage areas (Schuring, mall overflow, and wetland outflow).
[b] Assumed (see Section 6.2.1).
[c] Year 2002 includes component volumes not measured in previous years.

The unknown component identified in Tables 6-1 and 6-2 is calculated by subtracting contributions from all known sources from the total Consolidated Drain values. Flow volume of this unknown component consists of contributions from direct rainfall in the channel, supplemental fountain flow, drain channel groundwater input, and other direct channel inputs (e.g., remaining direct connections from nearby roads and neighborhoods) that could not be directly measured.

To understand stormwater thermal impacts associated with both the pre- and post-construction Consolidated Drain watershed contributions, it is necessary to first separate storm flow (i.e., wet weather flow) in the Schuring Drain from its baseflow (i.e., dry weather flow).

Because of the high imperviousness and lack of stormwater treatment facilities in this adjacent sub-watershed, flow in the Schuring Drain experiences dramatic rises and falls during a typical storm event. Combined with a steady baseflow, this allowed for the separation of storm flow from baseflow by studying the hydrograph of this drain. It was determined that any flow ≤ 0.045 m^3/s in the Schuring Drain was considered a dry weather flow and any flow >0.045 m^3/s was considered wet weather flow.

Based on this assessment, Tables 6-3 and 6-4 list percent breakdowns (including component contributions during dry vs. wet weather in the Schuring Drain) for volume and thermal additions during the post-construction two-month comparative period.

Table 6-3. Percent Volume Contribution to Portage Creek during the Two-month Comparative Period of August-September 2002 for the Post-construction Consolidated Drain Watershed (as Monitored in the Consolidated Drain Channel just Upstream of its Confluence with Portage Creek).

Contribution	Dry weather	Wet weather	Treated	Total
Wetland Out	-	-	21.9	21.9
Mall Overflow	-	1.1	-	1.1
Schuring	51.3	7.7	-	58.9
Other [a]	?	?	-	18.1

[a] Unaccounted load contributions to the Consolidated Drain channel calculated by total minus measured contributing drainage areas (Schuring, mall overflow, and wetland outflow).
? Cannot be distinguished.

Table 6-4. Percent Heat Contribution to Portage Creek during the Two-month Comparative Period of August-September 2002 for the Post-construction Consolidated Drain Watershed (as Monitored in the Consolidated Drain Channel just Upstream of its Confluence with Portage Creek).

Contribution	Dry weather	Wet weather	Treated	*Total*
Wetland Out	-	-	27.1	27.1
Mall Overflow	-	1.5	-	1.5
Schuring	45.9	8.8	-	54.7
Other [a]	?	?	-	16.7

[a] Unaccounted load contributions to the Consolidated Drain channel calculated by total minus measured contributing drainage areas (Schuring, mall overflow and wetland outflow).
? Cannot be distinguished.

Using Tables 6-1 through 6-4, a number of general observations for pre- to post-construction periods can be made. First, the effect of the treatment system on thermal loading is evident when the loads to Portage Creek from the Consolidated Drain in 1999 and 2002 are compared (Table 6-2). Despite more rainfall in the post-construction years, less thermal load was added to the creek once the load from the Schuring Drain was excluded (for 2002). Second, reduction of volume and thermal load within the treatment system in the comparative period of August-September in 2002 over pre-construction conditions (Tables 6-1 and 6-2) was consistent with the 42-day post-construction modeling period examined in Chapter 5.0, where the treatment system was noted to reduce stormwater volume and thermal load by 44% and 47%, respectively. Third, the newly introduced flow from Schuring Drain is the most dominant of the post-construction components draining to Portage Creek (Tables 6-1 and 6-2). The bulk of the Schuring Drain flow is composed of dry weather groundwater contribution (Tables 6-3 and 6-4).

The implications of these results are two-fold. First, construction of the stormwater treatment facility reduced thermal loading to Portage Creek from the Consolidated Drain watershed primarily by reducing runoff volume (see also Chapter 5.0). Second, the dominant contribution of dry weather baseflow at groundwater temperature from Schuring Drain (and to a lesser extent the Consolidated Drain) illustrates the opportunity to utilize existing coldwater baseflows to buffer the temperature regime of treated stormwater discharging from the wetland at air temperature.

6.3 Flow-weighted Thermal Loads in the Post-construction (2002) Consolidated Drain Watershed

A flow-weighted temperature index is developed here as a method of locating the most thermally enriching areas in a watershed based on the comprehensive dataset of flow and temperature for key hydrological components of the watershed in the summer months of 2002. The utility of developing such an index is introduced first, followed by details of the calculation and application of this index in the Consolidated Drain watershed.

6.3.1 Temperature and Thermal Load

Using Equation 5.4.2-2 to calculate thermal loads results in a heat content parameter expressed as an energy unit (i.e., calories). The value of this parameter is an extensive property influenced strongly by the flow rate (Q). Temperature, however, is an intensive property directly connoting the degree of thermal pollution. As shown earlier, the stormwater treatment system, with the addition of a wetland to the traditional detention ponds, was capable of substantially reducing stormwater thermal load (over 40%) during the dry-to-dry post-construction modeling period in 2002. This occurred primarily through a flow reduction mechanism—infiltration and, to a lesser degree, shading and evapotranspiration. However, average daily temperature analysis showed that despite the significant thermal load reduction, the treatment system could only reduce the wetland outflow temperature to that of ambient air. Because Portage Creek, as a coldwater stream, had a lower-than-air temperature (Figure 5-4), the treatment system outflow was still a thermal pollution source.

In evaluating the thermal loading contributions of various flow components in the post-construction Consolidated Drain watershed, the energy flux approach, while yielding important information on the relative distribution of thermal loads, could not pinpoint the sources and sinks of thermal pollution in the system. Except for the flow within the treatment system, there were large fluctuations in flow rate for other flow components in the watershed (e.g., runoff from the Crossroads Mall is zero during dry weather conditions). As a result, average temperature comparison was not suitable in identifying thermal sources or sinks particularly over an extended period, such as the two-month comparative period of August-September. It was therefore necessary to find a parameter that reflected the polluting potency of every thermal source over an extended period while taking into account the flow factor. A parameter termed the temperature equivalent was developed to accurately reflect the effectiveness of the treatment system in mitigating thermal load and to evaluate the thermal pollution contributions in the drainage area and the treatment system.

6.3.2 Temperature Equivalent

For this study, the following equation is used to define temperature equivalent (TE):

$$TE = \frac{\left(\frac{\sum Q_i \cdot \rho \cdot T_i \cdot C \cdot t_i}{\sum Q_i \cdot t_i}\right)}{\rho \cdot C} = \frac{\sum Q_i \cdot T_i \cdot t_i}{\sum Q_i \cdot t_i} \qquad (6.3.2\text{-}1)$$

where: TE = temperature equivalent (°C);
Q = flow rate (m^3/s);
ρ = water density (assumed to be constant at 1×10^3 kg/m^3);
T = water temperature (°C);
C = water heat capacity (assumed to be constant at 1 cal/g/°C); and
t = time interval (s).

The subscript *i* denotes the segment in the time series used in recording the values of the corresponding parameters.

The result of Equation 6.3.2-1, TE, has units of degrees Celsius. It is the flow-weighted average temperature of the flow component calculated in a pre-defined period of time (August 1 through October 1, 2002); i.e., a first moment approach to weighted averaging. In common pollutant terms, TE is analogous to event mean concentrations (EMCs). As denoted by Strecker et al. (2001), comparing the statistics of the long-term EMCs of the inflow and outflow is a technique far more scientifically sound than the conventional approach of averaging the load reduction efficiency of all the storm events monitored for evaluating stormwater treatment system effectiveness. The conventional approach essentially treats all storms as equal while storms vary widely in runoff volumes produced and associated pollutant concentrations. By using TE, each increment of temperature change of a flow component in the watershed is weighted against the total volume of this component over the two-month comparative period. The resulting TE value is an indicator of the thermal polluting potency of each flow component, which takes into account flow volume. TE is a function of the physical characteristics of the flow source area.

A simple schematic of the treatment system in Figure 6-1 shows the TE value for each of the flow components in the 2002 post-construction Consolidated Drain watershed during the August-September comparative period.

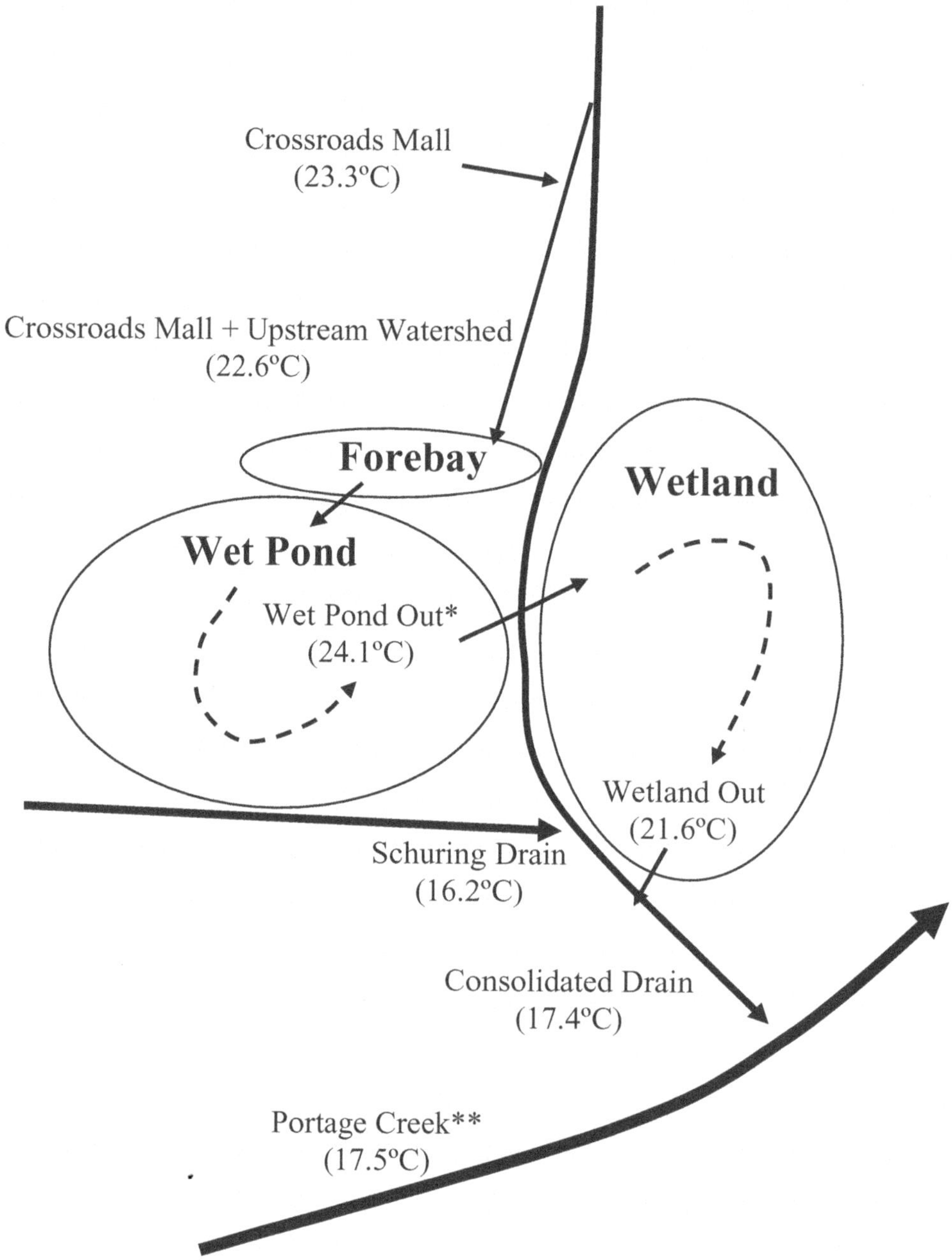

Figure 6-1. TE of Flow Components in the Post-construction Consolidated Drain Watershed during the Two-month Period of August 1, 2002, to October 1, 2002. Average air temperature during the same time period was 20°C. *Data unavailable at the wet pond out location from 15:10 9/6/02 through 10:00 9/27/02 due to thermistor malfunction [N=40 days monitored]. **Data unavailable in Portage Creek for 4 days in August due to USGS flow gage malfunction [N=57 days monitored].

TE was also calculated for Portage Creek upstream of its confluence with the Consolidated Drain channel using U.S. Geological Survey (USGS) gage station data and temperature readings from a nearby thermistor installed in the creek. The average air temperature for the period was 20°C.

Stormwater runoff from the 30.8-ha regional shopping center at the heart of the urban corridor (i.e., the Crossroads Mall) had a TE value of 23.3°C. This was 3.3°C higher than the average air temperature and 5.8°C above Portage Creek TE. The combined mall and upstream urban watershed runoff had a TE of 22.6°C. This TE pattern indicates that impervious surfaces in the mall and its upstream areas were absorbing heat and subsequently releasing stored heat to stormwater runoff. Without a treatment system, runoff from these two areas was a direct source of thermal pollution to the receiving stream.

With a prolonged detention period in the sediment forebay and wet detention pond, the TE value of the stormwater rose to 24.1°C, an indication of the warming effect of these open ponds. The wet pond volume was designed for a minimum 24-hour detention time for the runoff produced by a two-year, 24-hour storm event. The detention time was effectively much longer as its outflow feeding the wetland was restricted to obtain a five to 10 day retention time in the wetland, which maintained standing water at depths below 0.31 m. The wetland, in contrast, was able to cool the outflow from the wet pond by 2.5°C, to a TE of 21.6°C. Overall, the treatment system lowered the temperature of collected stormwater by 1°C TE, with the wetland cell accounting for all the cooling. However, with regard to the receiving stream, the treatment system is still considered a source of thermal pollution as its outflow TE was 4.1°C higher than Portage Creek. This in-system TE pattern confirms the observation made earlier regarding the effectiveness of each treatment cell in mitigating or exacerbating thermal pollution of urban stormwater runoff. This also substantiates that the wetland's maximum cooling capabilities are tied to ambient air.

The other two flow components in the post-construction Consolidated Drain watershed are from the Schuring Drain and the Consolidated Drain channel starting from its confluence with Schuring and ending at its discharge to Portage Creek (see Figure 6-1). The location and channel bed elevation are such that both of these two drains are fed by a steady groundwater input that has a lower-than-air temperature. Although both drain channels receive pulses of stormwater due to the unmitigated stormwater runoff in the Schuring Drain watershed, the Schuring Drain contribution was largely groundwater-fed baseflow. As a result, the TE value of the Schuring Drain flow was the lowest among all the flow components during the two-month period in 2002, including that of Portage Creek. This means that the Schuring Drain (i.e., its baseflow) generally constituted a thermal pollution mitigating source to the wetland outflow. This was particularly significant during dry weather conditions when the wetland continued to discharge to the Consolidated Drain.

Groundwater baseflow of the lower portions of the Consolidated Drain channel also served to offset the warmer wetland outflow. Immediately above its confluence with Portage Creek, the Consolidated Drain TE value was 17.4°C, slightly lower than the creek. This suggests the post-construction Consolidated Drain watershed over the two-month comparative period in 2002 did not add thermal pollution to Portage Creek on a flow-weighted thermal load basis. This combination of features illustrates the successful mitigation of thermal loading to Portage Creek.

When calculating and interpreting TE values in a watershed, use caution when choosing the time span for basing TE values. If a long time span is used, such as the TE values calculated here, TE represents the long-term thermal polluting potency of each flow component in a watershed. As a result, flows composed mostly of steady low-temperature baseflow show less or no chronic thermal impact on the receiving stream. If a short time period is chosen, such as the duration of a rain event, TE reflects the acute thermal impact of various flow components from a watershed during that event. The advantage of using a long time span, however, is that it can capture the effect of thermal enrichment/mitigation processes that are of a continuous and long-term nature, such as those taking place in stormwater treatment structures.

6.4 Thermal Regime Changes in Portage Creek

The treatment system was designed primarily to achieve the mitigation of stormwater impacts on Portage Creek. Part of the effectiveness investigation was to understand how the treatment system changed the thermal regime in Portage Creek. Knowing how flow components from the watershed contribute differently to temperature variation in the creek provides some understanding of the complexity of this question. Because multiple flows converge in the Consolidated Drain channel before it meets Portage Creek, it is necessary to consider all contributions when explaining thermal regime changes observed in this study. In this section, thermal regime changes in Portage Creek after the construction of the treatment system are examined from two perspectives: overall temperature changes and daily temperature fluctuations.

6.4.1 Temperature Changes

The thermal regime change in Portage Creek induced by the stormwater treatment system and the reconfiguration of the Consolidated Drain channel flow components in 2001 can best be represented by the water temperature differences among select thermistors near the drain and creek confluence. These include thermistors T3 (Consolidated Drain channel ~15 m upstream of its confluence with Portage Creek [Figure B-1]), T4 (Portage Creek ~5 m upstream of its confluence with the Consolidated Drain), and T5 (Portage Creek ~50 m downstream of the confluence). Thermistor T3 measures water temperature of the Consolidated Drain channel just before it meets the creek; T4 measures water temperature of the creek just before it receives water from the Consolidated Drain channel; and data from T5 directly reflects the temperature change in the creek as a result of the mixing of water from the Consolidated Drain channel and the creek.

Figure 6-2 shows temperature differences between T3 and T4, and between T5 and T4, along with the rainfall amount recorded at the five-minute interval over the two-month comparative period of August-September 2000. Figure 6-3 shows the same parameters for 2001. Year 1999 produces a similar graph to 2000 (not shown), both of which represent pre-construction conditions. The advantage of using T4 minus T3 and T4 minus T5 for comparison instead of individual thermistor readings is that with T4 as the reference point, year-to-year climatic variations (e.g., temperature and precipitation) can be eliminated. Year 2002 data are not available due to the loss of T4 to vandalism.

Three notable trends can be discerned from Figures 6-2 and 6-3. First, both graphs indicate that the Consolidated Drain channel had a lower water temperature than the creek during periods of dry weather (T4 - T3 >0).

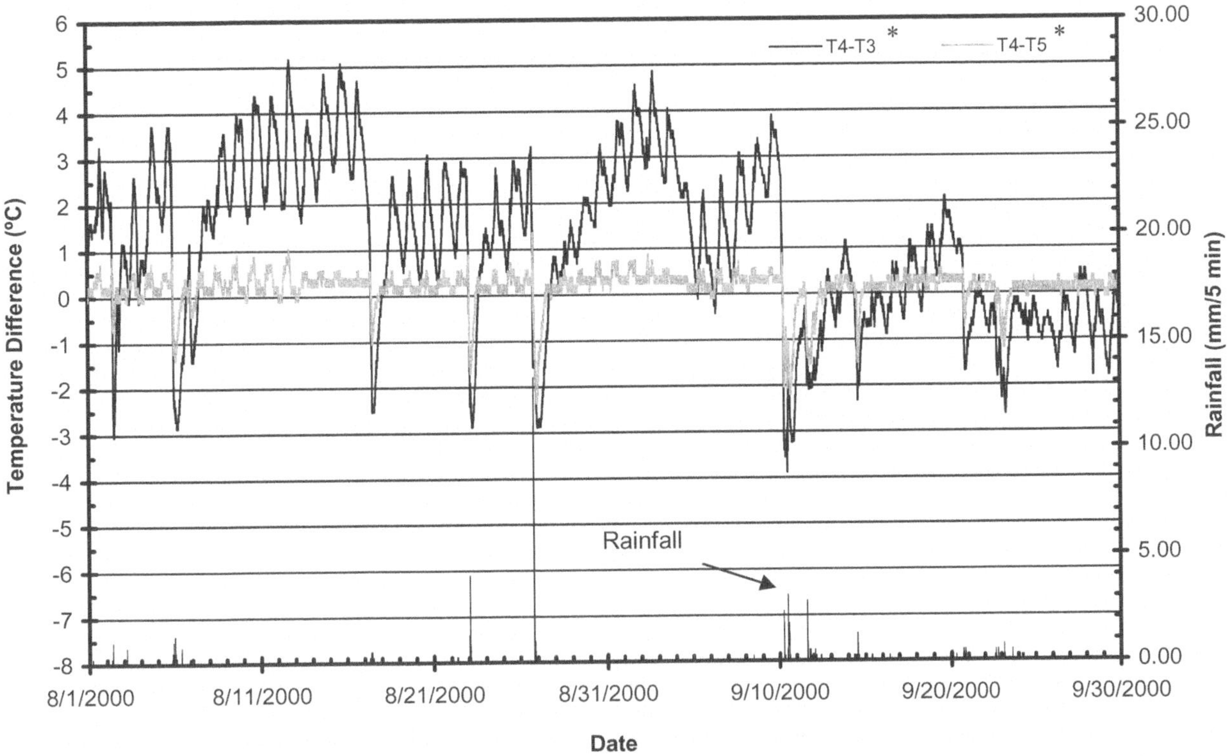

Figure 6-2. Water Temperature Differences Between the Consolidated Drain Channel and Portage Creek in August and September 2000 (Before the Stormwater Treatment System Came Online). *Temperature difference between thermistors T4 (at Portage Creek ~5 m upstream of the confluence with the Consolidated Drain) and T3 (at Consolidated Drain ~15 m upstream of the confluence), and between thermistors T4 and T5 (at Portage Creek ~50 m downstream of the confluence with the Consolidated Drain).

When there was no surface runoff, the Consolidated Drain channel was primarily fed by groundwater, both before and after the construction of the treatment system. The magnitude of the difference, however, was larger before the treatment system was in operation in 2000 (Figure 6-2) than after the system was functioning in 2001 (Figure 6-3). After the treatment system came online in late July 2001, there was a steady release of detained stormwater runoff by the system through the wetland outflow. The detained water had temperatures near ambient air temperature, which increased the water temperature of the normally groundwater-fed Consolidated Drain channel.

Second, during dry periods in 2000, water in the creek upstream of the Consolidated Drain confluence had a slightly higher temperature than that downstream of the confluence (T4 - T5 > 0). This is likely caused by cooling of the creek water by the Consolidated Drain groundwater baseflows. In contrast, during dry periods in 2001, T4 and T5 readings were quite similar with only minimal differences (T4 - T5 ≤ 0) due to the release of detained higher-temperature stormwater runoff by the treatment system.

Third, sharp downward spikes of both the T4-T3 and T4-T5 lines are evident in Figure 6-2 during rain events, indicating higher temperature readings from T3 and T5. This is a sign of the flashy thermal impact of the Consolidated Drain watershed on the creek. Stormwater runoff heated by impervious surfaces in the watershed became the major component of the

Consolidated Drain flow and elevated Consolidated Drain water temperatures up to 4°C higher than the creek. As a result, T5 readings were elevated. In 2001, although these downward

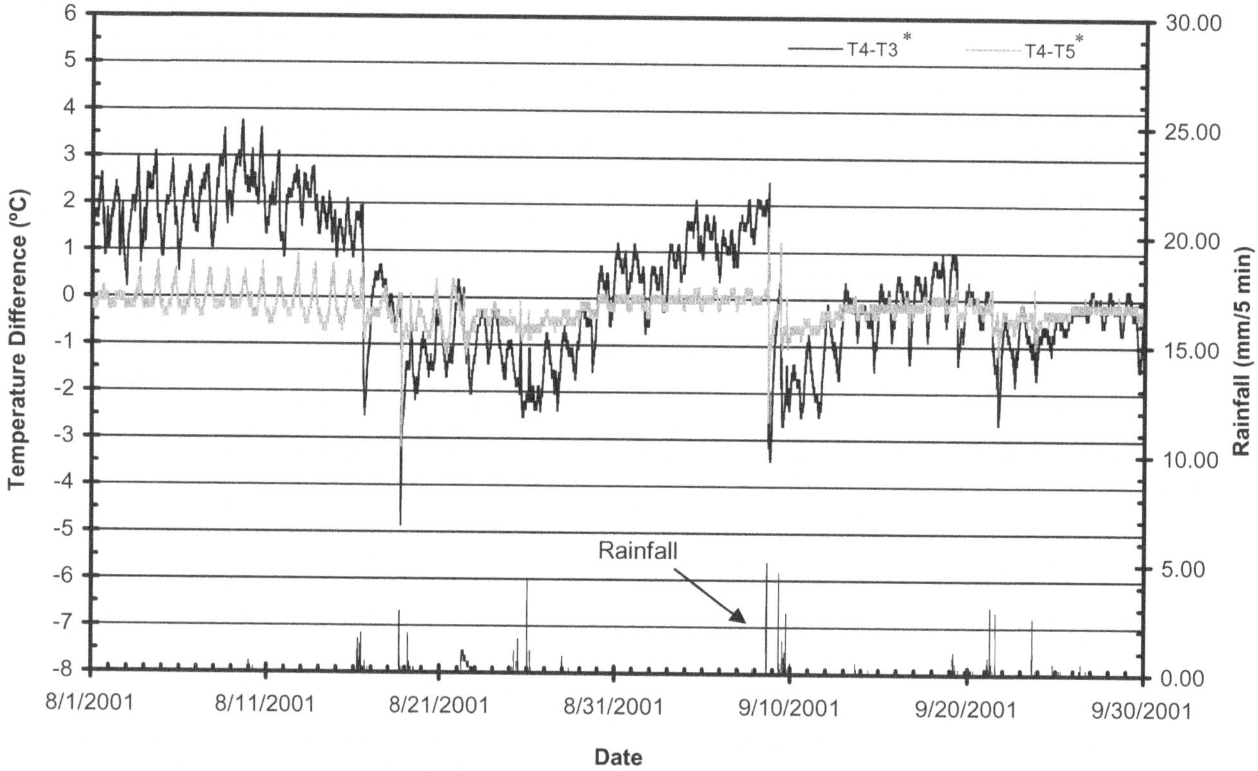

Figure 6-3. Water Temperature Differences Between the Consolidated Drain Channel and Portage Creek in August and September 2001 (After the Stormwater Treatment System Came Online). *Temperature difference between thermistors T4 and T3, and between thermistors T4 and T5; see Figure 6-2 for thermistors T3, T4, and T5 locations.

spikes still existed, they were smaller in magnitude, particularly for T4 - T5. This change can be explained by the detention of stormwater runoff in the treatment system after it came online in late July 2001. Because of the treatment system and the connection of the Schuring Drain to the Consolidated Drain channel, stormwater runoff from the Schuring watershed became the major flow component in wet weather conditions after July 2001. This caused the downward T4 - T3 spikes in Figure 6-3. However, these spikes in general were small compared to those in Figure 6-2, reflecting the fact that storm runoff from the Schuring watershed is smaller in volume compared to that originating from the post-construction Consolidated Drain watershed. The only exception was the rain event taking place between 15:50 and 17:35 on August 18, 2001. This event (12.4 mm) was intense in the beginning (5.84 mm for the first 15 minutes) and happened during the warmest time of the day. In Figure 6-3, the decrease of T4 - T5 values after a rain event was not as dramatic as that in Figure 6-2, another indication of the lower flow volume in the post-construction Consolidated Drain channel during wet weather conditions.

Statistical analysis of these annual data further illustrates the temperature regime change in Portage Creek as a result of the construction of the Consolidated Drain stormwater treatment system. In the post-construction months of 2001, the mean value of T4-T5 became negative,

Table 6-5. Statistics of the Five-minute Water Temperature Difference (ΔT = T4 - T5) in Portage Creek Upstream (T4) and Downstream (T5) of the Confluence with the Consolidated Drain Channel for August and September in 1999, 2000, and 2001.

	1999	2000	2001
N of cases	17,272	17,281	17,281
Minimum (°C)	-3.361	-2.350	-3.180
Maximum (°C)	3.667	1.361	1.530
Range (°C)	7.028	3.711	4.710
Median (°C)	0.067	0.222	-0.240
Mean (°C)	0.057	0.157	-0.226
95% CI upper[a]	0.062	0.163	-0.222
95% CI lower[a]	0.051	0.152	-0.230
Variance	0.118	0.137	0.088
Skewness	-4.051	-2.967	-0.568
Kurtosis	26.775	11.849	4.705

[a] 95% confidence interval upper and lower limits.

indicating an overall temperature increase from T4 to T5 due to the higher post-construction Consolidated Drain temperatures (T3), as noted in Table 6-5. The lower variance value for 2001 suggests that because of stormwater detention, there were fewer sharp temperature increases in the creek. The closer to zero skewness value and the lower kurtosis value in 2001 implies that the data distribution about the mean is less skewed towards higher T5 values. This distribution is closer to a normal distribution than the distributions in 1999 and 2000 shown in Figure 6-4. In summary, statistics in Table 6-5 confirmed that with the treatment system in operation, the acute stormwater thermal loading to the creek was diminished, while there was a slight but prolonged temperature increase caused by the slow release of stormwater detained in the treatment system. It is uncertain whether this small increase (0.24°C on average) was ecologically significant, however, it is not believed to be that dramatic given the urban setting of this stream.

In 2002, due to the loss of thermistor T4 during the August-September period, it was not possible to perform the same analysis as above. Instead, thermistor T6, located about 1000 m upstream of T4, was used to compare the temperature regime change in the creek caused by stormwater runoff from the Consolidated Drain and Schuring watersheds in 2002. Because of the data gap in 2000 (no T6 data for August-September 2000), data from 1999, 2001, and 2002 were statistically analyzed for T6 - T5. Figures 6-5 and 6-6, and Table 6-6 demonstrate that under post-construction conditions (2001 and 2002), the acute stormwater thermal shock to Portage Creek was diminished. This is reflected by the statistical distribution of T6 - T5 data about the mean being less skewed. This distribution is also closer to a normal distribution than in 1999 (pre-construction).

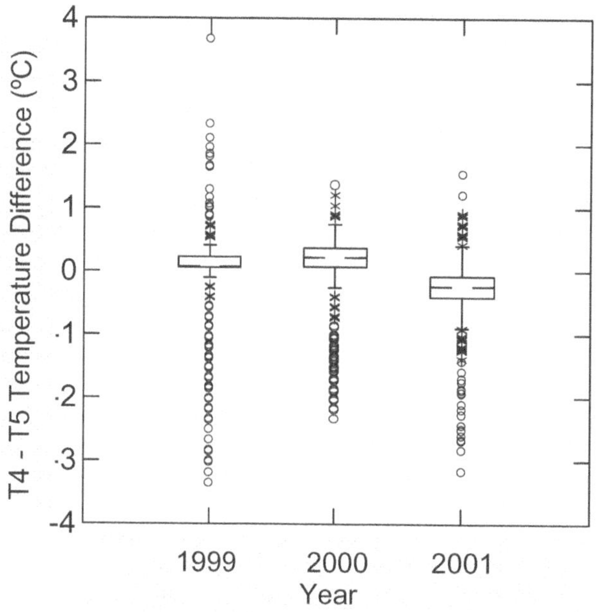

Figure 6-4. Box Plot of the Water Temperature Difference in Portage Creek Upstream (T4) and Downstream (T5) of the Confluence with the Consolidated Drain for August and September 1999, 2000, and 2001.

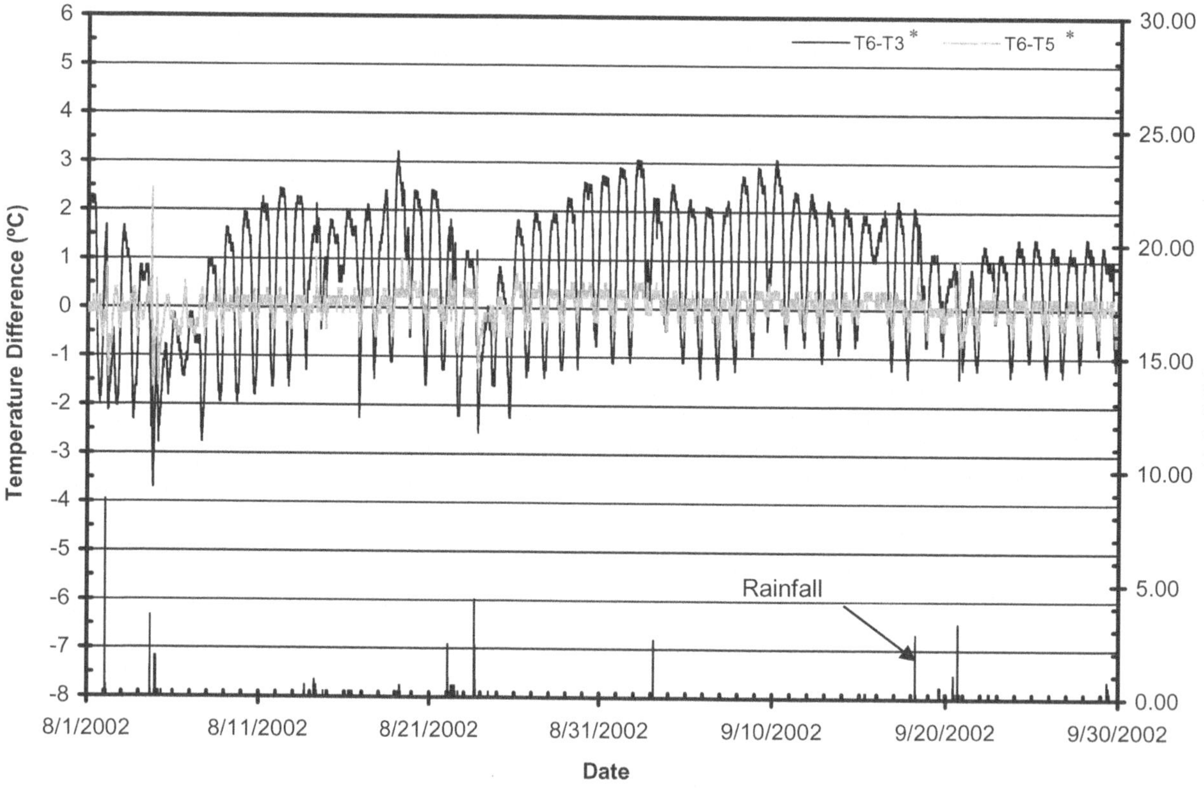

Figure 6-5. Water Temperature Differences Between Consolidated Drain Channel and Portage Creek in August and September 2002 (First Full Season with Full Riparian Growth After the Stormwater Treatment System Came Online).
*Temperature difference between thermistors T6 (at Portage Creek ~1000 m upstream of the confluence with the Consolidated Drain) and T3, and between thermistors T6 and T5.

WERF

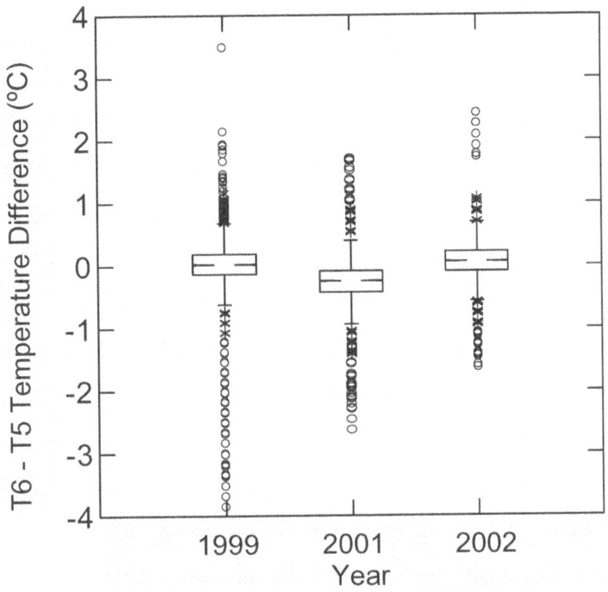

Figure 6-6. Box Plot of the Water Temperature Difference in Portage Creek ~1000 m Upstream (T6) and Downstream (T5) of the Confluence with the Consolidated Drain for August and September in 1999, 2001, and 2002.

Table 6-6. Statistics of the Five-minute Water Temperature Difference (ΔT = T6 - T5) in Portage Creek ~1000 m Upstream (T6) and Downstream (T5) of the Confluence with the Consolidated Drain Channel for August and September in 1999, 2001, and 2002.

	1999	2001	2002
N of cases	17,270	17,281	17,278
Minimum (°C)	-3.861	-2.630	-1.630
Maximum (°C)	3.489	1.710	2.430
Range (°C)	7.350	4.340	4.060
Median (°C)	0.022	-0.250	0.060
Mean (°C)	-0.030	-0.233	0.024
95% CI upper[a]	-0.023	-0.227	0.028
95% CI lower[a]	-0.036	-0.239	0.020
Variance	0.193	0.153	0.076
Skewness	-2.341	0.019	0.019
Kurtosis	15.055	2.758	3.488

[a] 95% confidence interval upper and lower limits.

Heating of the open water by solar radiation went undeterred resulting in the prolonged small temperature increase in the creek (Figure 6-3). In 2002, emerging vegetative cover produced by the growth of plant populations (cottonwood, willow, cattail, and various colonizing grasses and weeds) provided greater shading. This resulted in an overall decrease of water temperature in the creek. The cooling effect suggested by Table 6-6 and Figure 6-6 is consistent with the analysis in Section 6.3.2 regarding the thermal contribution from the Consolidated Drain channel to Portage Creek using the temperature equivalent concept.

6.4.2 Daily Temperature Fluctuations

To illustrate daily temperature fluctuations in the system, Figures 6-7, 6-8, and 6-9 show the difference between the daily temperature maximums and minimums for thermistors T3, T4/T6, and T5 for the two-month comparative period in 2000, 2001, and 2002, respectively. Daily air temperature fluctuations are also plotted for 2001 and 2002 (data are not available for 2000 due to instrument malfunction).

Figures 6-8 and 6-9 indicate that daily temperature fluctuations in the Consolidated Drain channel and Portage Creek closely followed those of ambient air, with exceptions occurring mostly during rain days. This suggests that despite substantial groundwater input in the Consolidated Drain channel below the Schuring Drain and the creek, heat transferred via the air (long-wave radiation and conduction) and direct short-wave solar radiation still caused substantial daily temperature fluctuations in the Consolidated Drain and the creek. Comparing Figure 6-7 with Figure 6-8 it is clear, however, that the construction of the treatment system increased daily temperature fluctuations in the Consolidated Drain channel (T3) relative to the creek, both upstream (T4) and downstream (T5) of the Consolidated Drain channel. Again, different degrees of channel shading were the key to this change. Before the construction of the treatment system, the Consolidated Drain channel was covered by a dense layer of vegetation growing in and around the channel. The Schuring Drain was not connected to the Consolidated Drain. Therefore, the Consolidated Drain channel during dry weather carried a small baseflow shaded almost completely by vegetation. This condition resulted in low daily temperature fluctuations in the Consolidated Drain channel during dry weather (Figure 6-7). After the

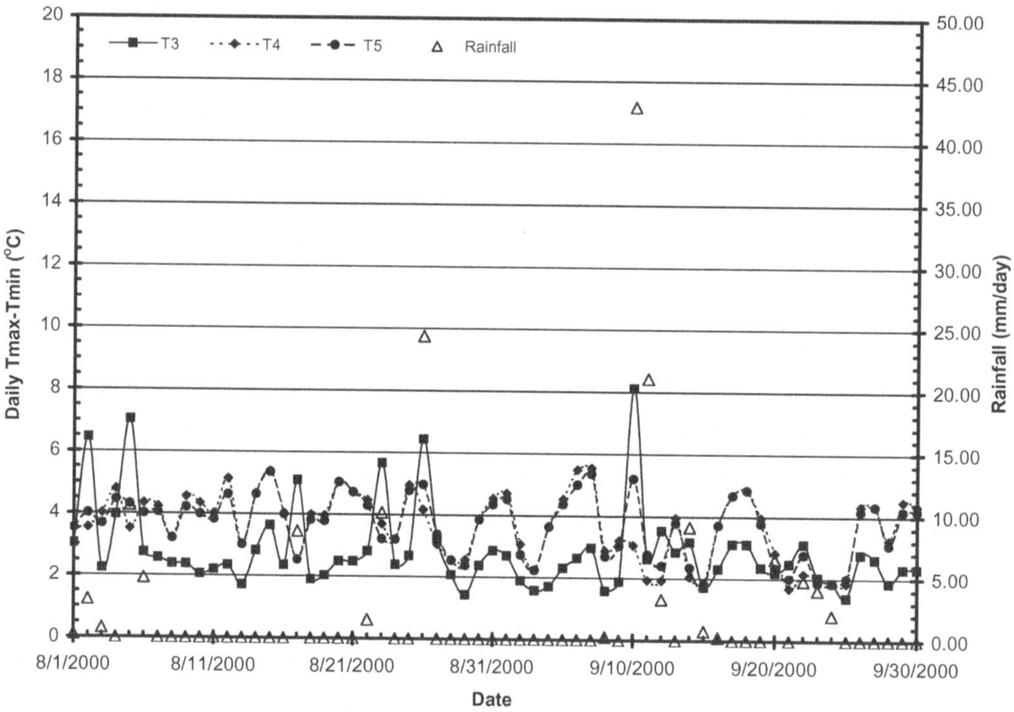

Figure 6-7. Daily Temperature Fluctuations of the Consolidated Drain Channel and Portage Creek in August and September 2000 (Before the Stormwater Treatment System Came Online), Illustrated by the Difference Between the Daily Temperature Maximum and Minimum.

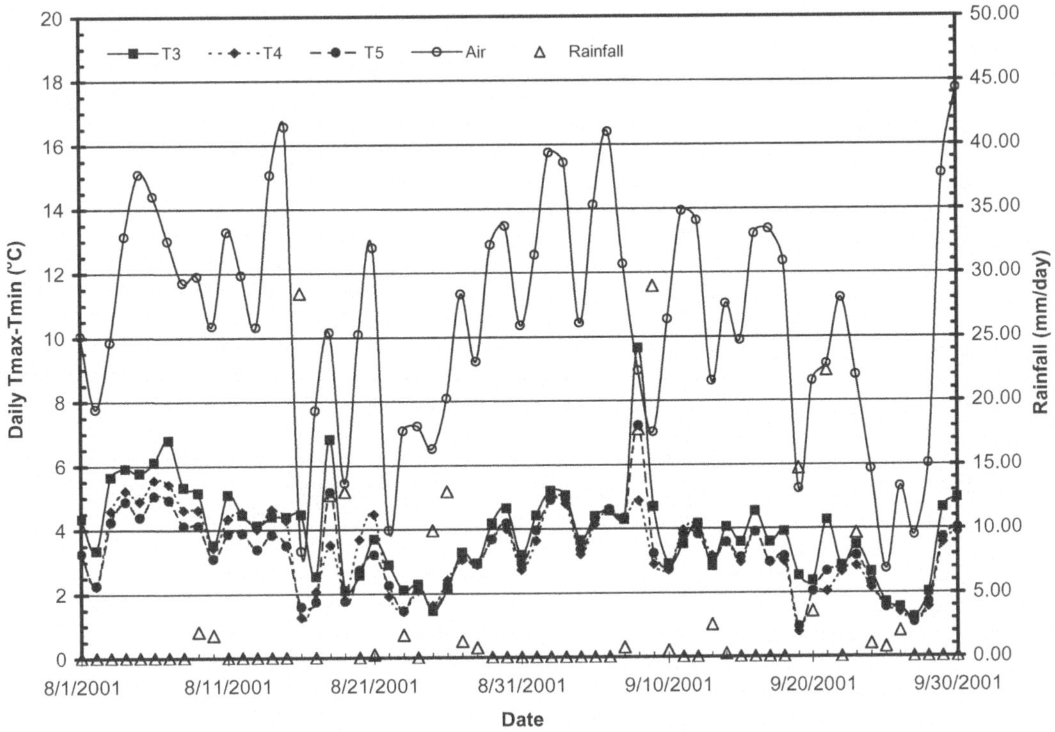

Figure 6-8. Daily Temperature Fluctuations of the Consolidated Drain Channel and Portage Creek in August and September 2001 (After the Stormwater Treatment System Came Online), Illustrated by the Difference Between the Daily Temperature Maximum and Minimum.

construction of the system, due to the channel reshaping and the introduction of the Schuring Drain, vegetation was lost and more flow was exposed to directly solar radiation. In addition, stormwater runoff from the Schuring Drain watershed is not mitigated and treated, resulting in flashy flows in the Schuring Drain during rain events. Consequently, flow in the Consolidated Drain channel experienced higher daily temperature fluctuations than the creek (T4 and T5). Nevertheless, with the treatment cells mitigating stormwater runoff, fewer peaks of high daily temperature fluctuation were observed in 2001 than 2000 (Figures 6-7 and 6-8) during wet weather, even though there were more rain days and greater total rainfall in 2001 than in 2000 during the comparative period (277 mm vs. 160 mm, respectively; Table 6-2).

More evidence of the effect of riparian and instream vegetative shading on stream temperature regime comes from the 2002 plot of daily temperature fluctuations of T3, T6, and T5 (Figure 6-9). As vegetation planted around the Consolidated Drain channel started to mature in 2002, the T3 daily temperature fluctuations would have been expected to return to the 2000 pattern. However, Figure 6-9 shows T3 had daily temperature ranges frequently exceeding 7°C (average 6.1°C). As a comparison, 2000 and 2001 had daily temperature ranges mostly below 6°C and averages of 2.8 °C and 3.9°C, respectively. The most likely reason for this surge of daily variation in the Consolidated Drain channel in 2002 is the additional flow from the fountain of the reflecting pond entering the open Consolidated Drain channel north of treatment system in the summer of 2002 (Figure 3-2). This additional source provided a continuous flow during daytime (the fountain did not operate at night) in the channel upstream of its confluence with the Schuring Drain. As described in Section 4.3.6, due to the lack of riparian shading on the part of the Consolidated Drain channel that connects the reflecting pond outlet and the mouth of the

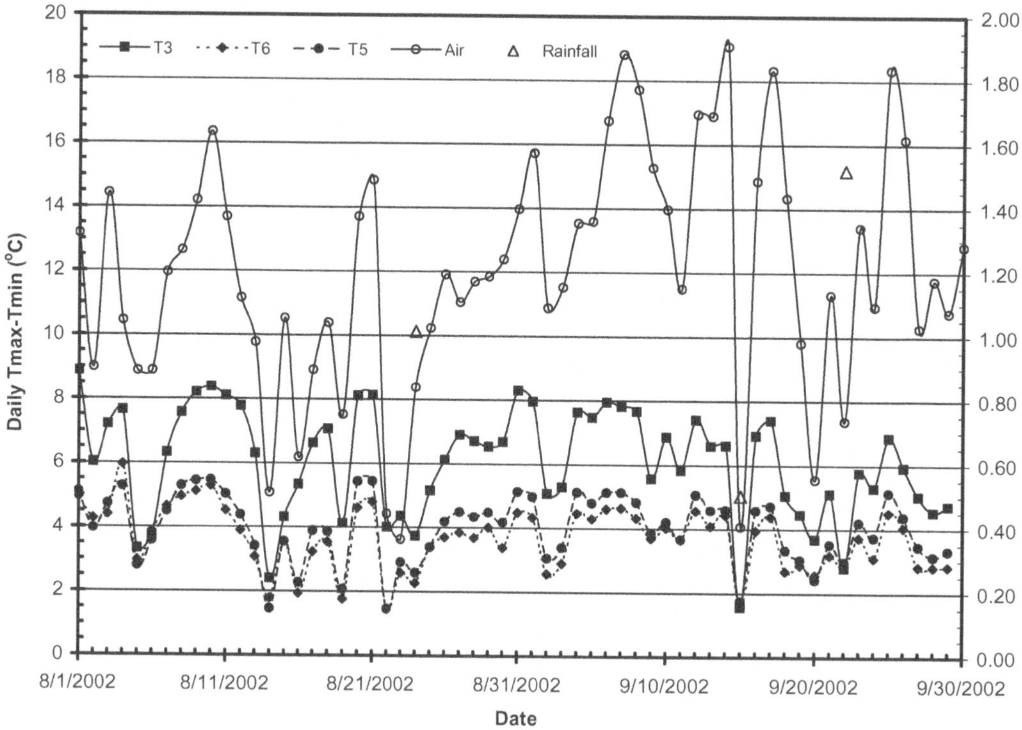

Figure 6-9. Daily Temperature Fluctuations of the Consolidated Drain Channel and Portage Creek in August and September 2002 (First Full Season with Full Riparian Growth After the Stormwater Treatment System Came Online), Illustrated by the Difference Between the Daily Temperature Maximum and Minimum.

Schuring Drain, water at the end of this part of the channel can be 5-10°C higher than the air during daytime (Figure 4-7). Therefore, this influx of warm water (~0.028 m^3/s) during the day can substantially increase the maximum daily temperature of the Consolidated Drain. As a result, thermistor T3 reported very high daily fluctuations in 2002.

A closer examination of the five-minute temperature record of thermistors T3, T4, and T5 for a period of 11 days in 2000 (Figure 6-10) and 2001 (Figure 6-11) further identifies the importance of shading in stream temperature regime regulation. Figure 6-10 shows that during dry weather when the Consolidated Drain channel was completely covered by vegetation, the daily water temperature minimum occurred at about 07:00 and 08:00 and the maximum occurred at about 17:00 and 18:00. This timing is about one hour later than the corresponding air temperature minimum and maximum. For T4 and T5, daily minimum and maximum readings took place at least one hour later (two hours later than the air). When the vegetation was not present in 2001, T3 had the same timing for daily minimum and maximum readings as T4 and T5 (Figure 6-11). This change can be explained by the following. For streams with extensive vegetative shading, water temperature is driven primarily by the surrounding air. Due to its high heat capacity, water temperature lags behind the rise and fall of the air. For streams exposed to direct sunlight, temperature will continue to rise even after air temperature peaks, as long as there is still solar radiation. As a result, the Consolidated Drain channel had daily temperature highs and lows lagging the air but preceding Portage Creek in 2000, which was only partially shaded (Figure 6-10). In 2001, however, the shading condition was similar in the Consolidated Drain and the creek. Consequently, readings of T3, T4, and T5 showed parallel daily variation patterns.

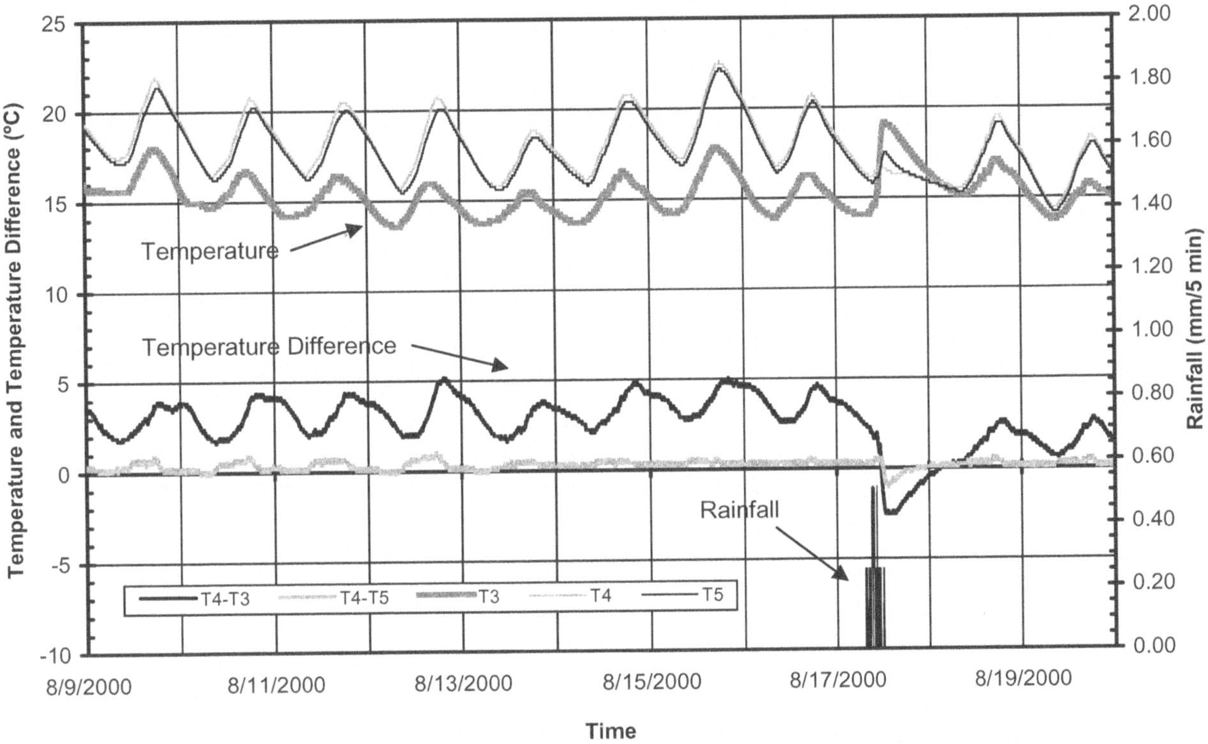

Figure 6-10. An Example of Five-minute Interval Temperature Records of the Consolidated Drain Channel and Portage Creek in August and September 2000 (First Full Season with Full Riparian Growth After the Stormwater Treatment System Came Online), Illustrated by the Difference Between the Daily Temperature Maximum and Minimum.

Figures 6-10 and 6-11 also show the temperature differences between the creek and the Consolidated Drain. Maximum and minimum differences between T4 and T3 in 2000 generally corresponded to the peaks and valleys of T4. In 2001, due to the parallel daily variation of T4 and T3, there was no clear pattern to be distinguished.

Overall, Figures 6-7 through 6-11 indicate that daily fluctuations of instream temperature are primarily a result of ambient air temperature change but can be exacerbated by storm events. The highest single daily temperature fluctuations (daily maximum minus daily minimum) for the Consolidated Drain (T3: 9.6°C) and Portage Creek (T5: 7.2°C) occurred during the two-month comparative period in 2000 and 2001. Both were recorded on September 8, 2001, when a one-hour 17.8 mm mid-afternoon rain event occurred. This is most likely due to the unmitigated runoff carried by the Schuring Drain. During dry weather, however, vegetative cover played a significant role in regulating daily stream temperature variations in the Consolidated Drain. Figures 6-7 to 6-11 also show that the Consolidated Drain could increase the daily fluctuations in Portage Creek up to 2.4°C (September 8, 2001) during wet weather. However, due to its relatively small flow rate (0.028 m^3/s to 0.057 m^3/s in the drain vs. 0.43 m^3/s to 0.45 m^3/s in the creek), the Consolidated Drain did not have much effect on the daily fluctuations of the creek during dry weather. This was apparent even when the wetland treatment cell discharged nearly continuously into the drain throughout the two month comparative period in 2001.

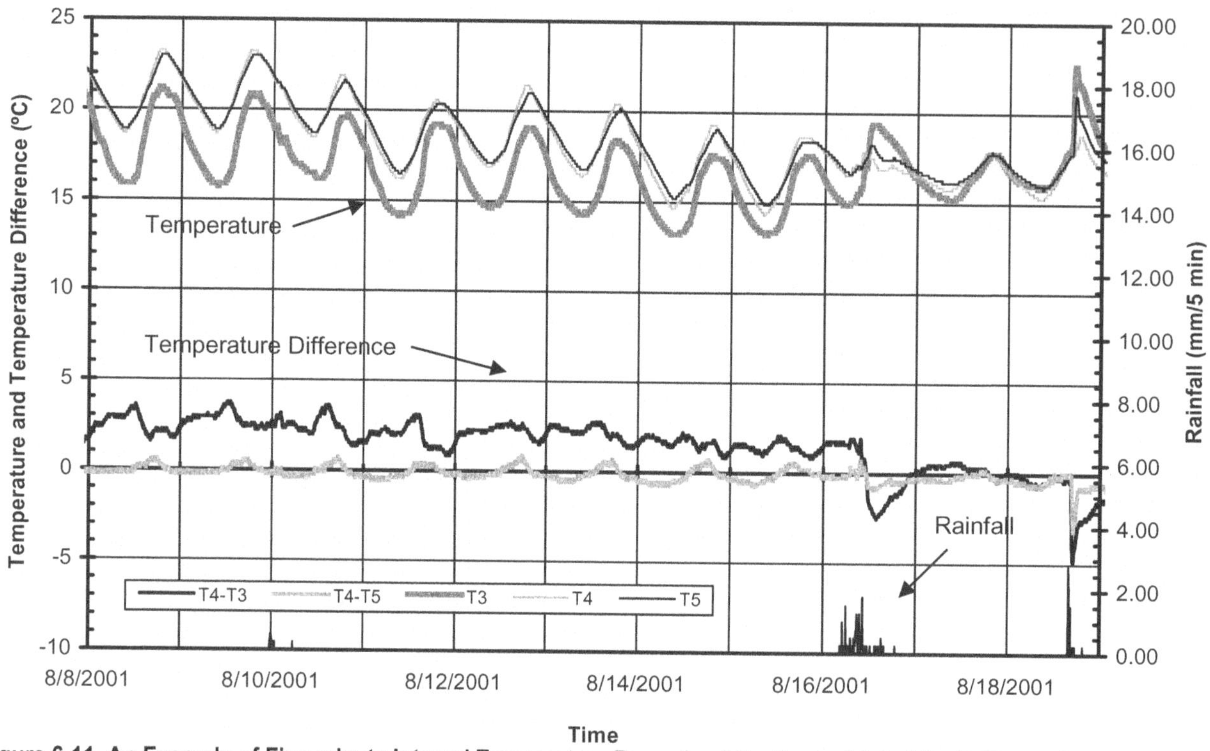

Figure 6-11. An Example of Five-minute Interval Temperature Records of the Consolidated Drain Channel and Portage Creek in August and September 2001 (First Full Season with Full Riparian Growth After the Stormwater Treatment System Came Online), Illustrated by the Difference Between the Daily Temperature Maximum and Minimum.

CHAPTER 7.0

MODELING OF STORMWATER THERMAL ENRICHMENT AT THE CROSSROADS MALL

This chapter describes preliminary efforts to build a thermal enrichment model for direct runoff temperature simulation in the Consolidated Drain watershed. The model covers principal heat transfer processes such as radiation, evaporation (sensible heat), and convection. It also includes physical drainage system components such as impervious pavement, open channels, closed conduits, confluences, ponds, and wetlands.

This chapter starts with an introduction to the model, while Appendix D provides background theory. Locations simulated by the model are described in Appendix B. The model is applied to the Consolidated Drain watershed, principally the 30.8-ha Crossroads Mall, for simulating two afternoon storm events. A sensitivity analysis is also presented. Conclusions from model simulations in terms of model applicability and key model parameters are provided.

7.1 Heat Transfer Model (HEATRAN)

Based on previous work (Van Buren, 1999; Haq, 2001; Picksley and Deletic, 1999; Verspagen, 1995; and Xie, 1993), a model called HEATRAN was constructed that computes runoff temperature directly, using a distributed heat balance for most components of a drainage. HEATRAN works with PCSWMM 2002. The central idea of the model is the combined heat balance and mass balance. The model covers principal heat transfer processes for an open water body: convection, rainfall mixing, evaporation (sensible heat), radiation, and conduction.

All situations or hydraulic elements involved in the watershed are generalized into a single description and applied to four broadly similar hydraulic situations or elements: 1) overland flow; 2) closed conduit or pipe flow; 3) open channel flow; and 4) open storages (ponds and wetlands). HEATRAN discretizes a watershed into the four categories of hydraulic elements and connects them as a network. Therefore, it is applicable to an infinitely large and complex drainage network system.

The energy budget is analyzed for every flow element in a drainage. There are two essential thermal processes: heat mixing in runoff water, and conduction inside the underlying body. For heat conduction on the ground, a finite difference method is employed to obtain a numerical solution. The model successfully combines both processes.

The convective heat-transfer coefficient between runoff and the ground surface is one of the major uncertainties in this research. Its determination is a difficult task in many cases. Dimensional analysis is employed to derive general expressions of the convection heat-transfer coefficients for different hydraulic elements and weather conditions given in Table 7-1. The expressions include a few adjustable coefficients which are calibrated against field data.

Table 7-1 Convective Heat Transfer Coefficients.

Hydraulic Elements	Dry Periods	Wet Periods
Closed conduits	Ignored	$h = F1 \cdot \frac{k}{R} \cdot \Pr^{F3}(\mathrm{Re}_R{}^{F4} + F2)$
Channels	$h = F5 \frac{ku}{\nu}(\Pr^{F7} + F6)$	
Subcatchment		$h = G1 \cdot \frac{k}{R} \cdot \Pr^{F3}(\mathrm{Re}_R{}^{F4} + F2)$ $G1 = e^{(F1 + F11 \cdot R_h + F14 \cdot \lvert Tr - Tg \rvert \cdot i^{F16})}$
Ponds and wetlands		$h = F1' \frac{k}{d}(\Pr^{F3'} + F2')$

where:

h = Convection heat-transfer coefficient, $\frac{watts}{m^2 \cdot {}^\circ C}$;

k = thermal conductivity, $\frac{J}{kg \cdot {}^\circ C}$;
Rh = fluid hydraulic radius, m;
d = water depth of pond, m;
Pr = Prandtl number;
Re_R = hydraulic radius-based Reynolds number;
U = flow velocity, m/s;
ν = fluid kinematic viscosity, m^2/s
Tr = runoff temperature, °C;
Tg = ground surface temperature, °C;
F1~F16 = calibration coefficients.

Programmed in MatLab language, HEATRAN comprises 18 subroutines. To handle the high sensitivity of low-flow waterbodies to temperature calculations, a special approach, called the step-calculation method, is employed to control the sensitivity and ensure continuity and stability of the results. The main principle is that heat transfer components are separately added and adjusted in the heat mixing calculation for runoff water. This method makes HEATRAN applicable to low flow.

To simulate temperature changes of runoff in an urban stormwater system, PCSWMM is used first to obtain a discharge hydrograph for all hydraulic elements—the RUNOFF module is used for subcatchments, TRANSPORT for closed conduits and channels, and STORAGE for pond elements. The hydrographs computed by PCSWMM are transferred to HEATRAN. Since an automatic data transfer is still unavailable, this step is completed manually using Microsoft Excel. Finally, HEATRAN is run with meteorological data to compute temperature changes separately for every element, and candidate parameters are calibrated to obtain a reasonable result.

7.2 Application to the Consolidated Drain Watershed

The total drainage area of the Crossroads Mall is 30.8 ha or 16.4% of the Consolidated Drain watershed (Figure 3-2). The mall area is divided into two drainage areas in the model: the east area (16 ha) which includes 11 pipes; and the west area (14.8 ha) with 12 pipes (K&A, 2002). Figure 7-1 gives the schematic layout of the pipes. Stormwater from the east and west areas eventually drains into manholes MH-10B and MH-10A, respectively. From MH-10A, runoff from the west area is conveyed by two parallel pipes to MH-10B, where the two flows are merged. The merged stormwater is delivered again by two parallel pipes to MH-9 (site 25, Appendix B), and then to overflow weir #2, which is connected to the Consolidated Drain via a 42-in. diameter underground pipe.

As for the upstream portion of the Consolidated Drain watershed from the overflow weir #2 (Site 10, Appendix B), surface runoff is collected by a pipe network. Eventually the stormwater is drained into a pipe at overflow weir #1. From the overflow weir #1, two conveyances deliver the stormwater. One is an open channel, the Consolidated Drain floodway channel, which is the upstream portion of Consolidated Drain. The other is the 42-in. diameter pipe, which passes under overflow weir 2 and reaches the sediment forebay. This portion of the stormwater flows through the forebay, the wet detention pond, the treatment wetlands, and finally into the Consolidated Drain channel. The latter accepts event volumes from smaller than two-year storms, and the former is a regular floodway (for event volumes greater than two-year) and baseflow conveyance.

Simulation efforts for the Consolidated Drain watershed are concentrated on summer storm events on the mall and adjacent areas upstream of the sediment forebay. Figure 7-1 schematically shows the discretization. In total, there are 51 elements: 24 subcatchments and 27 pipes. Among the pipes, 24 to 48 are concrete, while 50 and 51 are high-density polyethylene (HDPE). Flow into the sediment forebay is composed of two parts: one from the mall area (Part I) and the other from adjacent areas (Part II) draining into the Consolidated Drain upstream of its confluence with the mall drainage (Figure 7-1). All overland flow elements (subcatchments) are considered to be impervious (This assumption may be not realistic for the Part II area, but the area calibration will remedy this.) Since PCSWMM transport can simulate only dendritic

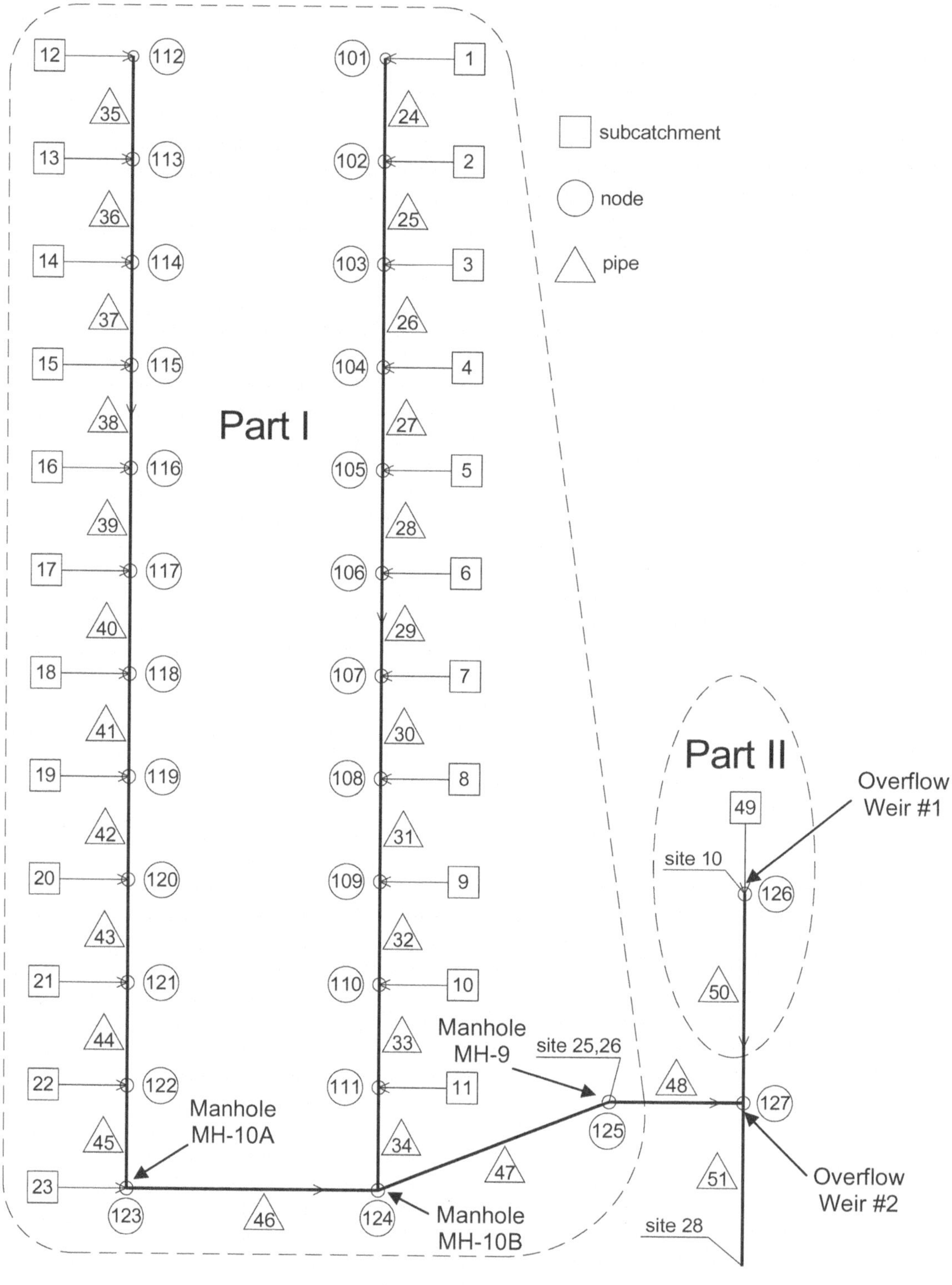

Figure 7-1. Discretization Scheme for the Crossroads Mall and Adjacent Areas.

Networks (James and James, 2000), the parallel pipes just below the mall bifurcation structure were combined into a single pipe (47, 48). The equivalent diameter of this single pipe is determined by flow rate calibration.

The simulation approach uses:

1. Flow data at mall runoff, site 25 and 26 (Appendix B) to calibrate the PCSWMM model of Part I, the Crossroads Mall property, and flow data at sediment forebay inflow, site 28, the sediment forebay inflow pipe, to calibrate the PCSWMM model of the combined drainage area of Part I and II; and thus obtains flow data for every hydraulic element in watershed.

2. Temperature data at mall runoff, site 25, 26 (Appendix B) to calibrate the HEATRAN model of Part I, the Crossroads Mall property; directly employs the temperature data at overflow weir 1, site 10, for runoff from Part II, the drainage area upstream of the mall; and uses temperature data at sediment forebay inflow, site 28 to calibrate the overall HEATRAN model output.

7.3 HEATRAN Simulation

Figures 7-2 and 7-3 show PCSWMM results are from the July 29, 2002, and August 13, 2002, respectively, events at the mall runoff (Site 25, 26; Appendix B) and sediment forebay inflow (Site 28; Appendix B) locations.

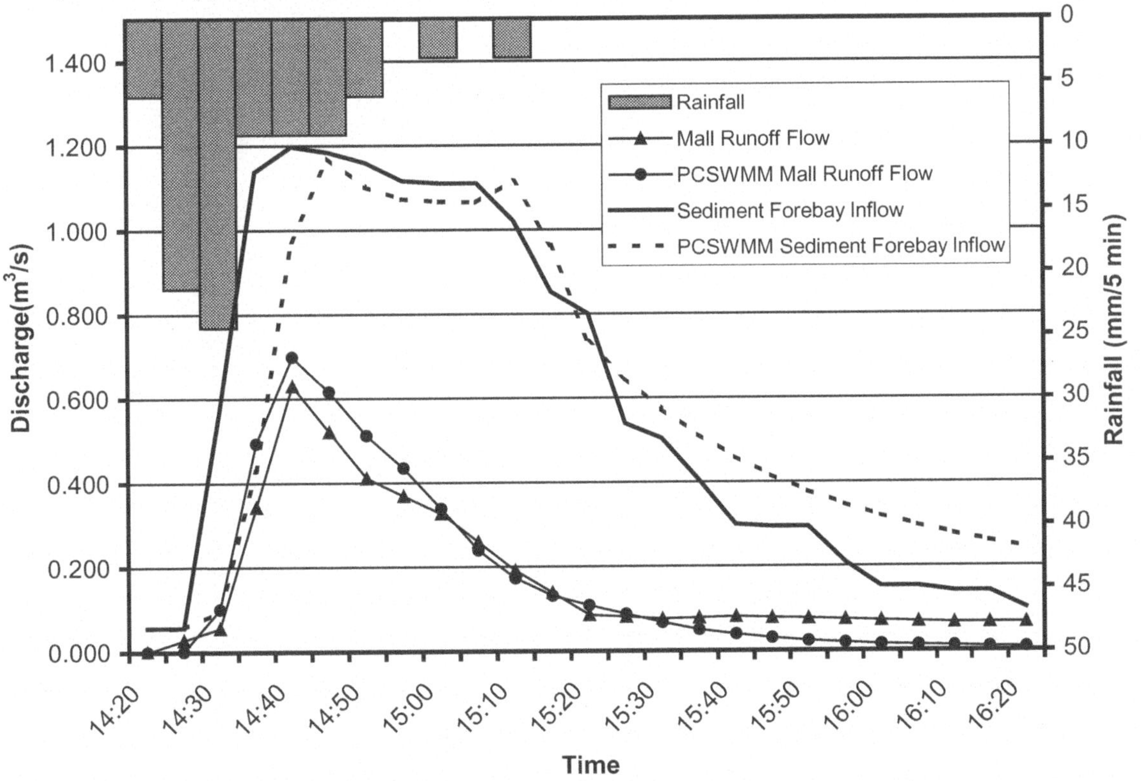

Figure 7-2. Discharge Hydrograph Observed and Modeled (PCSWMM) at Mall Runoff and Sediment Forebay Inflow Locations (July 29, 2002).

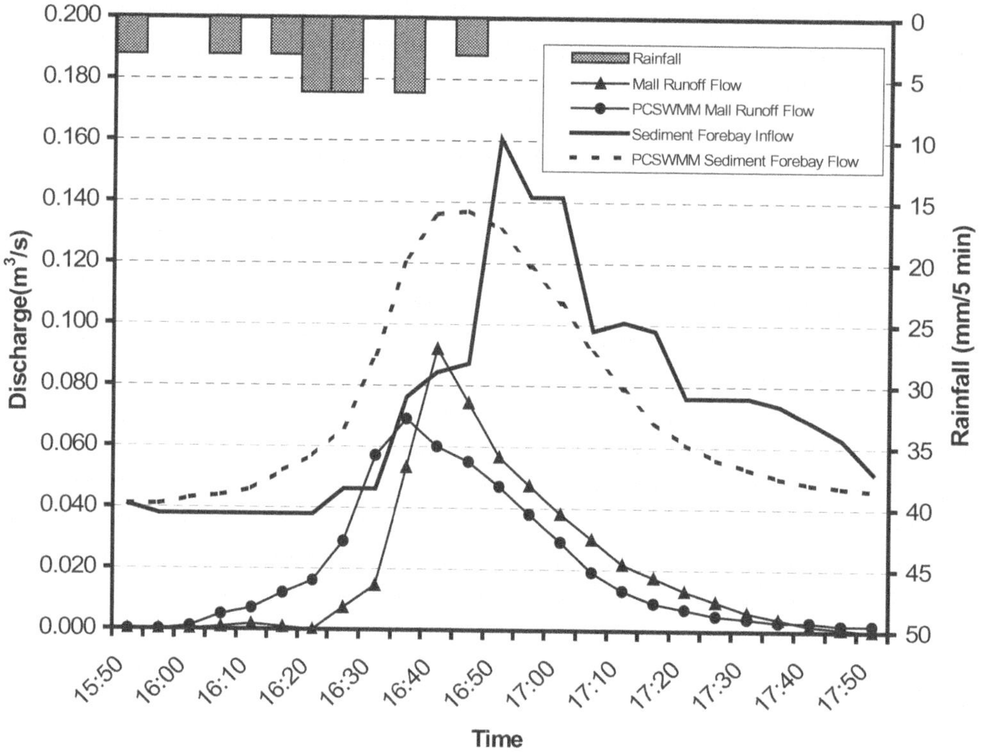

Figure 7-3. Discharge Hydrograph Observed and Modeled (PCSWMM) at Mall Runoff and Sediment Forebay Inflow Locations (August 13, 2002).

HEATRAN was run for the watershed system and the computed water temperatures are compared to the observed values in Figures 7-4 through 7-7 from the July 29, 2002, and August 13, 2002, events at the mall runoff (Site 25, 26; Appendix B) and sediment forebay inflow (Site 28; Appendix B) locations. The results basically agree with the observed data.

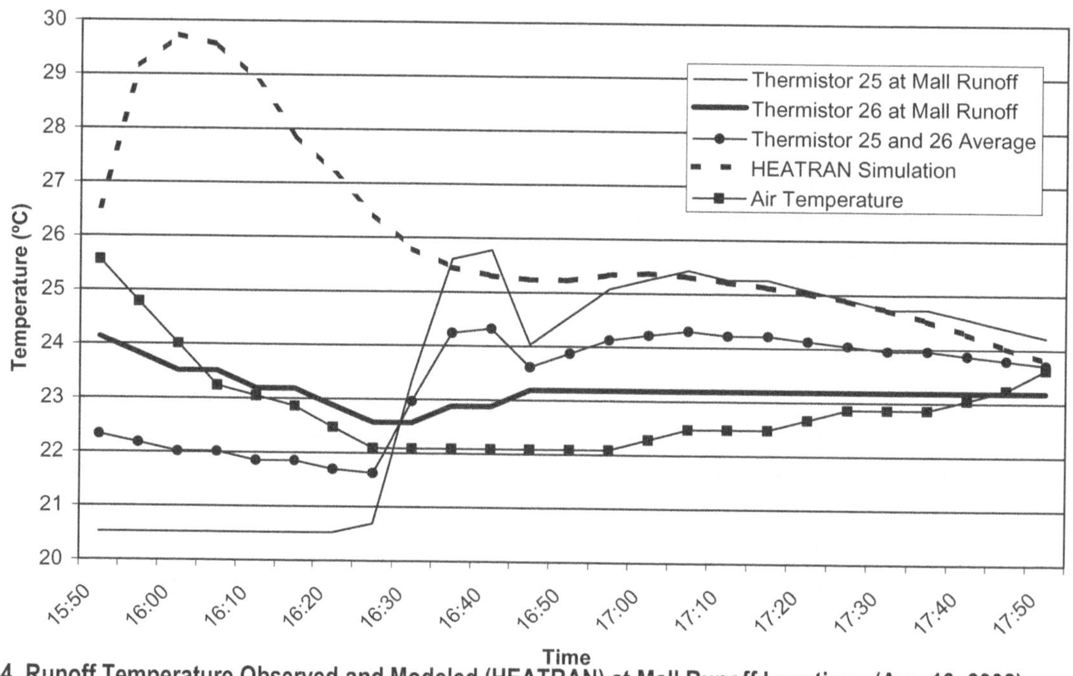

Figure 7-4. Runoff Temperature Observed and Modeled (HEATRAN) at Mall Runoff Locations (Aug 13, 2002).

WERF

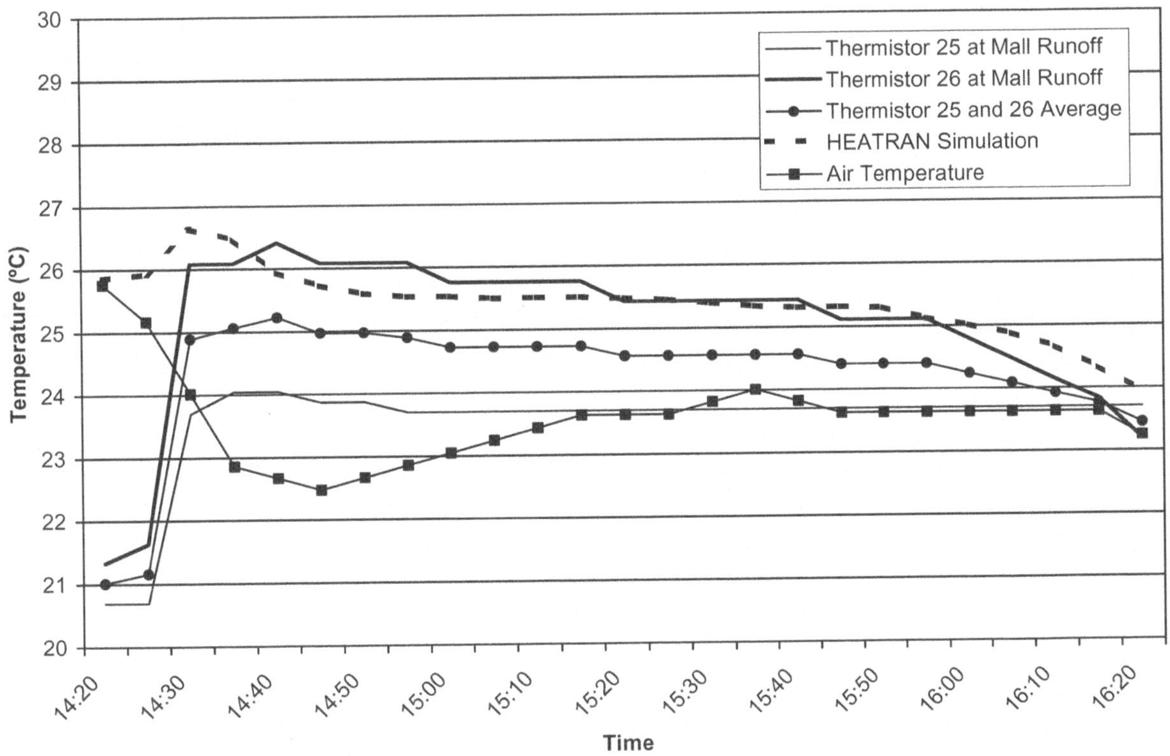

Figure 7-5. Runoff Temperature Observed and Modeled (HEATRAN) at Mall Runoff Location (July 29, 2002).

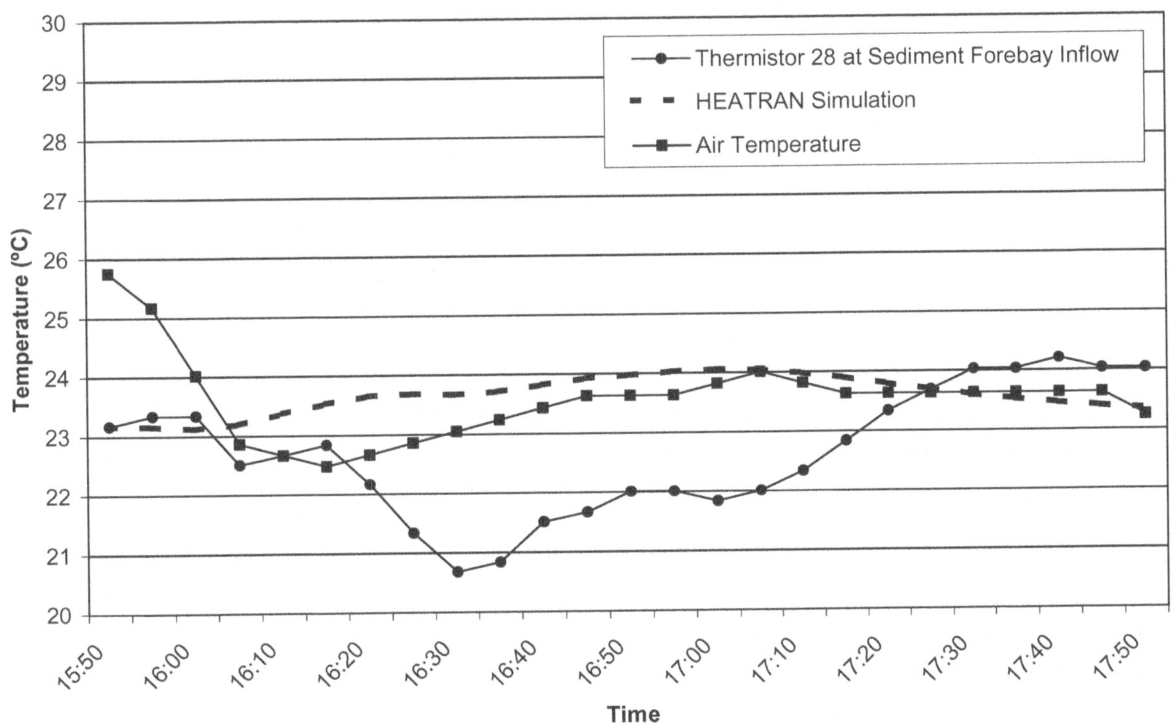

Figure 7-6. Runoff Temperature Observed and Modeled (HEATRAN) at Sediment Forebay Inflow Location (August 13, 2002).

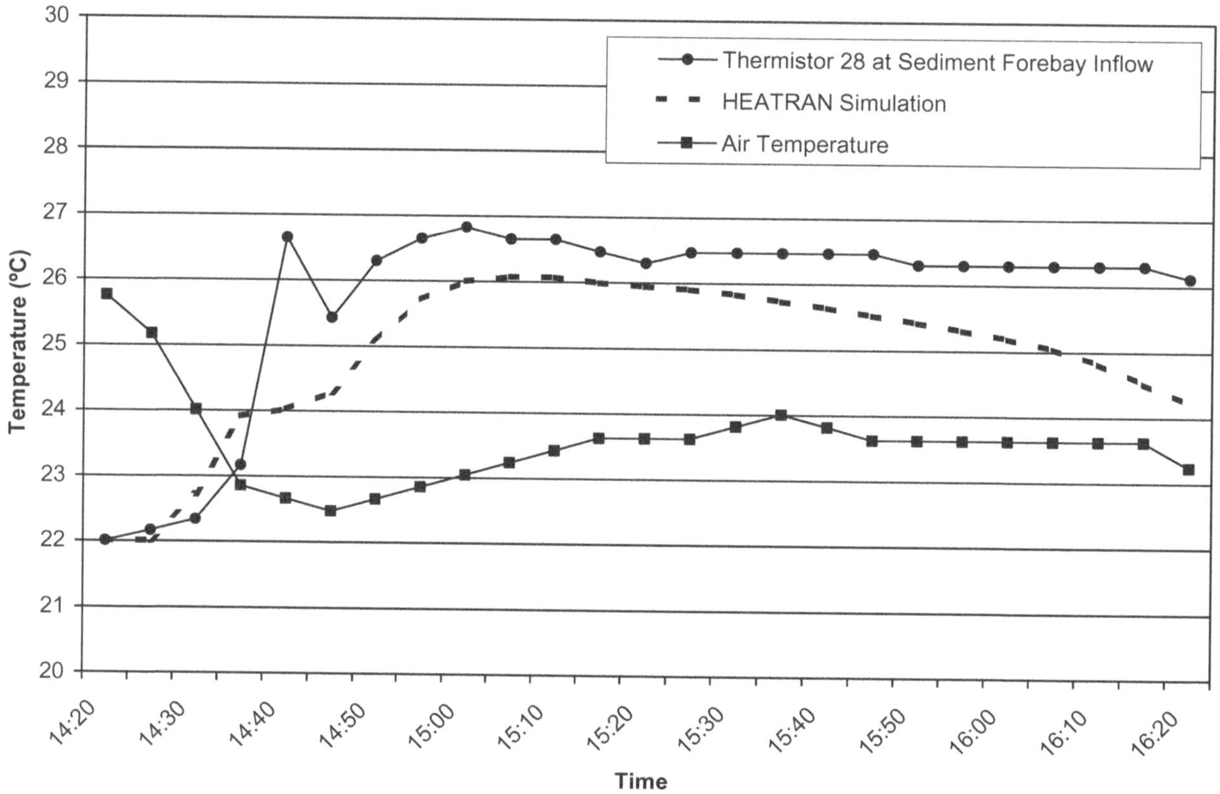

Figure 7-7. Runoff Temperature Observed and Modeled (HEATRAN) at Sediment Forebay Inflow Location (July 29, 2002).

Table 7-2 compares HEATRAN predictions of post-event surface temperatures in subcatchment 10 to observed readings and shows reasonable accuracy.

Table 7-2. Partial Surface Temperature Comparison for the Crossroads Mall.

	July 29, 2002	**August 13, 2002**
Period of rain event	14:20 - 16:20	15:50 - 17:50
Time of comparison	15:20 - 15:32	18:00
Measured surface temperature range (N=18)	28.2 °C - 30.2 °C	29.6 °C - 34.1 °C
Computed surface temperature[a]	28.6 °C - 28.7 °C	33.7 °C

[a] Computed surface temperature for subcatchment 10.

7.4 HEATRAN Sensitivity Analysis

HEATRAN contains numerous calibration parameters as shown in Table 7-3. In addition, many approximations were also made, such as pavement structure, pipe bedding structure, and initial temperature profiles beneath surfaces underlying runoff. Even the rainfall temperature is uncertain. It is difficult to comprehensively analyze the sensitivity of all the uncertain factors.

Hence, only several important coefficients were selected to be analyzed for sensitivity based on HEATRAN modeling experience.

Table 7-3. Calibrated Coefficients in the HEATRAN Model.

Terms	Remarks			Calibrated Value
Overall				
rra	Ratio of rainfall temperature to air temperature			1.0
alvar	Allowable varying range for temperature within a time step			±10%
mamo1~3	Deduction fraction in step-calculation			±10%
wei1	Weighted fraction for conduction boundary condition treatment			0.05
wei2	Weighted fraction for conduction boundary condition treatment			0.05
wei3	Weighted fraction for conduction boundary condition treatment			0.9
Subcatchment				
F1	Convection heat transfer coefficient	Constant term in G1 (wet period)		-13.5
F2		Constant term (wet period)		0
F3		Exponent of Prandtl number (wet period)		0
F4		Exponent of Reynolds number (wet period)		1
F5		Constant term (dry period)		1
F6		Constant term (dry period)		0
F7		Exponent of Prandtl number (dry period)		0.5
F11		Coefficient of flow depth in G1(wet period)		5000
F14		Coefficient of rainfall intensity term in G1 (wet period)		0.33
F16		Power of rainfall intensity in G1 (wet period)		1/10
F8	Coefficient of solar radiation (wet period)			0.001
F9	Coefficient of evaporation (wet period)			0.001
F10	Coefficient of solar radiation (dry period)			0.1
Pipe				
F1	Convection heat transfer coefficient	Constant term	Concrete pipe	0.001
			HDPE	0.00001
F2		Constant term		0
F3		Exponent of Prandtl number		0.4
F4		Exponent of Reynolds number		0.5
F12	Coefficient of evaporation	pipe 24 through pipe 48		0.9988
		pipe 50 and 51		1.0

Four parameters were suspected to dominate (most significantly influence the modeling result). They were: 1) convection heat transfer coefficient for subcatchments for wet periods (*chtcsw*, which involves the effects of *F1, F11, F14* and *F16*); 2) coefficient of solar radiation for subcatchments for wet periods (*F8*); 3) coefficient of evaporation for subcatchments for wet periods (*F9*); and 4) coefficient of evaporation for pipes (*F12*). To determine the sensitivity of these factors, the July 29, 2002, rain event was taken as the input because the rainfall amount of the August 13, 2002, event was small. The peak runoff temperature for sites 25, 26 at 14:30 was designated as the objective function. For most coefficients, increments were set between ±50%. Some coefficients may have a different increment due to computational stability criteria or other constraints. Figure 7-8 provides results.

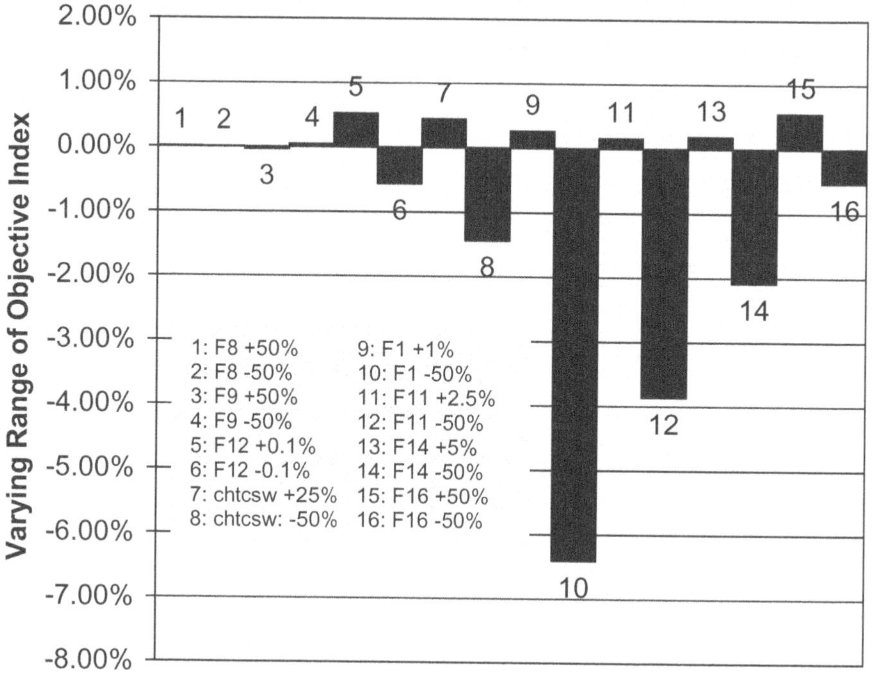

Figure 7-8. Sensitivity Analysis in Terms of Runoff Temperature at site 25, 26 (14:30, July 29, 2002) Expressed as Varying Range (%) of Objective Index.

Among the four parameters, evaporation inside pipes had the dominant effect on controlling the runoff temperature at a given location and time. With a minimum change (±0.1%), F12 changed the objective index by 0.5% and -0.6%. It is evidently important to consider evaporation inside pipes, though this process is poorly understood. Considering the relatively minor effects of evaporation and solar radiation, heat from the pavement is evidently the most important thermal source for overland sheet flow, except for the rainfall itself.

Another dominant parameter was *chtcsw*, the convection heat transfer coefficient (Table 7-1). Among components of *chtcsw F1*, *F11*, *F14*, and *F16* were all sensitive. *F11* represents the effects of flow depth on heat transfer for overland flow, while *F14* and *F16* represent the effect of rainfall intensity (Table 7-3). Radiation and evaporation in subcatchments were relatively insensitive. At 14:30 on July 29, 2002, rainfall intensity approached its peak value while radiation and evaporation became minor.

7.5 Findings from HEATRAN Modeling

As a deterministic and continuous temperature model, HEATRAN is reasonably efficient and computes temperature changes in a comprehensive way for a typical urban watershed.

However, a number of uncertainties make modeling difficult. Due to the lack of relative information, some parameters still have to be assumed, notably those for the underlying body of the modeled hydraulic elements. Considering these uncertainties, it is unlikely that the computed temperature will ever match the observed data.

However, the critical period for watershed management planning for runoff temperature is the low-flow, warm summer days. HEATRAN handles such situations properly. Several important outcomes can therefore be drawn based on this modeling effort:

- For subcatchment elements, rainfall characteristics (e.g., duration, amount, and temperature) and temperature difference between rainfall (or air) and the ground surface are the two most important controlling factors for heat transfer processes during wet periods, where the effect of radiation and evaporation is limited.
- For the convective heat transfer in the subcatchment element, rainfall intensity and flow depth have a heating or positive thermal effect enrichment effect.
- Apart from the convective heat transfer between runoff and the pipe wall, an extra heat loss exists inside pipes, related to flow depth.
- The pipe network has a cooling or negative thermal enrichment effect.
- Several convective heat transfer coefficients are recommended for further development or examination including the following (for description of symbols, see Table 7-1):

A) for pavement in wet periods,

$$h = \frac{k}{R_h} \cdot \mathrm{Re}_R \cdot e^{(-13.5 + 5000\, R_h + 0.33 \cdot |Tr - Tg| \cdot i^{0.1})}$$

B) for pavement in dry periods,

$$h = \frac{ku}{\nu} \mathrm{Pr}^{0.5}$$

C) for concrete closed conduits in wet periods,

$$h = 0.001 \cdot \frac{k}{R} \cdot \mathrm{Pr}^{0.4} \cdot \mathrm{Re}_R^{\ 0.5}$$

D) for HDPE closed conduits in wet periods,

$$h = 0.00001 \cdot \frac{k}{R} \cdot \mathrm{Pr}^{0.4} \cdot \mathrm{Re}_R^{\ 0.5}$$

Note that these equations are highly empirical. They may not be applicable to circumstances other than those at the Crossroads Mall, Portage, MI.

HEATRAN jointly calculates ground surface temperature and runoff temperature for stormwater drainage systems, and is reasonably efficient and comprehensive. It still requires copious data support and appropriate assumptions. The working principles of HEATRAN are very clear that all simulations are based on one single theme: the heat budget. It should be recognized that because this theme is simple, all parameters involved in the model are easy to understand and adjust.

CHAPTER 8.0

Conclusions and Implications

This chapter synthesizes previous chapters of the report and identifies critical issues associated with stormwater thermal enrichment and mitigating strategies. Where appropriate, section and chapter numbers are referenced.

8.1 Background

According to U.S. EPA (2004), there are 447 approved thermal modification TMDLs in the United States, of which 372 specifically list temperature as the pollutant. Most of these TMDLs were established in the Pacific Northwest and Mountain regions to protect coldwater habitats, particularly for salmonids. Wisconsin, Louisiana, and Georgia also have established temperature TMDLs. Increased direct solar radiation to streams caused by loss of riparian shading has been identified in most of these TMDLs as the primary nonpoint heat source. Point sources include power plants, wastewater treatment plants, and other industrial sources. Although mentioned in some TMDLs (e.g., Upper Klamath Lake Drainage TMDL in Oregon), warm surface stormwater runoff is not considered a major heat source, likely due to the predominant non-urban land use types in these temperature TMDL watersheds.

However, as areas of the nation continue to undergo land development and urbanization, the protection of coldwater streams from thermal pollution in current and potential urban/suburban areas has become an increasingly important issue in efforts to restore beneficial uses of these urban streams. This is particularly significant in the Midwest, where in summer months hot temperatures coincide with high precipitation characterized by frequent afternoon thunderstorms. Further, individual pollutant TMDLs and their corresponding remediation practices for urban impaired waters may ignore the thermal issue. This will lead to situations where these TMDLs have fully achieved their objectives of reducing the load of specific stressors but still do not meet the ultimate goal of pollution control (i.e., restoration of the ecological integrity and the beneficial uses of impaired waters).

Urban stormwater BMPs, for example, are traditionally designed to reduce pollutant loading and moderate flow. Typically, stormwater pollutants of most concern have been TSS, oil and grease, nutrients, pesticides, other organics, pathogens, heavy metals, and salts. With Phase II Stormwater Regulations that took effect in March 2003, communities are expected to spend millions of dollars for urban stormwater BMPs. Correspondingly, the role of these BMPs to achieve temperature TMDL goals is becoming more important than ever, particularly in urban areas with coldwater streams. Unfortunately, knowledge on this subject has been very limited with sporadic information suggesting that traditionally designed BMPs can increase the temperature of stormwater (Schueler, 1987; Galli, 1990; James and Verspagen, 1997; Van Buren et al., 2000a). Until this study, information and data that quantify the thermal impact of stormwater BMPs in a typical urban setting have been lacking. Moreover, design practices and new tools that can be used to mitigate these impacts have not been comprehensively researched and developed.

The discharge of thermally enriched stormwater to receiving waterbodies has been overlooked in research efforts aimed to abate the environmental impacts of urban runoff. Rainwater temperatures are usually assumed to be the same as ambient air temperature, yet ambient air temperatures are typically higher than coldwater stream temperatures. Thermal enrichment of stormwater runoff takes place when runoff acquires heat from constructed surfaces. Runoff in conveyances, channels, and detention ponds (online and offline) undergoes constant heat exchange with contact surfaces and ambient air. It also can absorb heat from direct solar radiation. This makes thermal pollution a much more challenging problem to address within watersheds that have coldwater streams.

Except for a handful of studies, most stormwater literature only acknowledges the importance of considering the temperature issue in designing stormwater treatment systems, but does not offer effective approaches to address it. Studies often fail to provide any meaningful quantification of the magnitude of the issue. This quantification is crucial for understanding the problem and the search for solutions. As reviewed in this study (Chapter 2.0), limited efforts have been attempted to measure or model all the critical aspects of heat exchange between runoff and the environment in a typical urban stormwater flow pathway. These critical aspects of heat exchange take place everywhere in the runoff pathway that starts from rainfall to overland flow, to conveyance pipes or channels, to treatment facilities, and eventually to the receiving stream. Consequently, there has not been a comprehensive analysis of the thermal issue. Results of previously available studies are often not readily applicable to the decision-making process in urban stormwater management.

This study represents the first attempt to fill this knowledge gap of stormwater thermal enrichment in urban settings. It provides a detailed characterization of the thermal enrichment process in each step of the stormwater flow pathway. The conditions and settings of this study were ideal for this characterization in that they represented a typical highly urbanized watershed with traditional and non-traditional stormwater treatment BMPs discharging to a coldwater receiving stream. Moreover, the system was monitored for more than four years, during which the treatment facilities (two detention ponds and a natural wetland) were designed and built according to the traditional standards for the purpose of flow detention and common pollutant removal. This provided the opportunity to quantify the thermal behavior of the system pre- and post-construction and investigate the effectiveness of traditional treatment approaches in mitigating thermal pollution. Results have important implications in the continued effort to protect and restore urban aquatic habitats, particularly coldwater streams.

8.2 Monitoring of Pavement and Runoff Temperature

Evidence from the pavement and runoff temperature monitoring for the Crossroads Mall drainage during July and August 2002 (Sections 4.3.1 through 4.3.4) indicated that pavement surface temperatures in unshaded areas were up to 17.6°C higher than shaded areas on sunny summer days. Surface temperatures reached as high as 52°C during the mid-afternoon hours. In addition, review of individual storm events revealed that while air temperature dropped during rain events, runoff temperature steadily increased, indicating that runoff was gaining heat despite decreasing air temperature. Maximum temperature differences between runoff and the air occurred during early morning hours and afternoon hours, indicating that runoff was heated from stored energy in the pavement during the day and night. Observations demonstrated that the Consolidated Drain watershed behaves in a manner consistent with the empirical understanding of urban stormwater thermal enrichment.

8.3 Thermal Behavior of Stormwater Treatment Facilities

During the August and September, average daily temperature of stormwater stored in traditional open water detention ponds can be nearly 7°C higher than ambient air temperature (Chapter 5). Chronic thermal enrichment of the receiving stream would likely have resulted even with a flow-moderated discharge of this water from the open ponds. Instead, a densely vegetated wetland, established as a third stage of treatment following traditional detention, brought the water temperature to near ambient air. This acclimation was attributed to volume reduction via infiltration during extended detention in the wetland, shading of the water from direct solar radiation, and some evapotranspiration. However, even with the treatment provided by the wetland, stormwater runoff still could not be returned to groundwater baseflow temperatures that define the thermal regime in the coldwater receiving stream.

Thermal load analysis performed on 11 individual storms monitored during a 42-day post-construction modeling period in 2002 revealed an overall heat reduction of 45% in the three treatment cells, with the wetland accounting for 97% of this reduction. Meanwhile, infiltration in the wetland supplied a 42% overall flow reduction within the treatment system. There was a strong correlation between flow reduction and heat reduction for the 11 events, suggesting that the primary heat load reduction mechanism was flow loss via infiltration. This strong relationship serves regions with similar climatic and land use characteristics by highlighting this critical design element for treatment of thermally enriched stormwater runoff where concern exists for coldwater receiving streams. Data analysis and linear regression (Figure 5-4) revealed that the system could achieve sizable heat load reductions even when the flow reduction was negative, implying other mechanisms were also involved. Shading provided by vegetation in the wetland is likely the most important element in this regard.

The implication of these findings is that open water detention ponds may not be a suitable option for stormwater treatment in protecting coldwater habitats. Well-vegetated wetlands and infiltration basins are better alternatives. Urban stormwater treatment efforts have traditionally focused on mimicking the pre-development flow regime and removing common pollutants from the runoff. Open water ponds may be effective devices to reach these goals. However, thermal load cannot be physically removed from the water, and heat exchange is a continuing process that does not stop with the cessation of rainstorms and runoff. These fundamental differences between thermal pollution and common constituent pollutants may render detention ponds as inappropriate options for terminal treatment of urban stormwater that is returned to coldwater receiving streams. This basic but critical finding supports infiltration practices rather than direct surface water discharges, a growing trend in stormwater management through source control.

8.4 Thermal Loading of Flow Components in the Watershed and Its Impacts on Stream Temperature Regime

Analysis of flow volume and thermal loading from various flow components monitored in this project indicated that dry weather output from the Consolidated Drain watershed is primarily composed of cold, groundwater-fed baseflow from both the Schuring Drain and the Consolidated Drain channel downstream of the Schuring Drain (Figure 3-2 and Section 6.2.3). The continuous output from the wetland contributed 22% of the total dry weather flows during the post-construction modeling period of 2002. This reflected the engineered storage and slow discharge of the treatment system. As the treatment wetland outflow maintained temperatures around ambient air, this flow component contributed a higher level of thermal loading (27%) relative to its flow contribution (22%), indicating a thermal loading source. However, the presence of the higher volume groundwater-fed baseflow (>50%) accounted for the majority of the cooling effect to the receiving stream.

Statistical analysis of temperature differences between Portage Creek upstream and downstream of its confluence with the Consolidated Drain channel demonstrated the acute thermal shock to the receiving stream from storms in pre-construction years (Section 6.4.1). In 2001, immediately after the treatment system became operational, similar analyses showed abatement of the acute thermal shock but pointed to a slight overall warming (0.22°C on average) of the receiving stream. This suggested chronic thermal enrichment as a result of the slowly released stormwater from the treatment wetland at ambient air temperatures. This was in combination with the exposed baseflows in the reconstructed upstream sections of the Consolidated Drain channel that had limited to no shading. In 2003, after vegetative maturation was progressing within the treatment system, particularly the establishment of vegetation along and within the Consolidated Drain channel, this chronic thermal enrichment was largely abated. Analysis showed only a negligible (0.07 °C on average) warming of Portage Creek by the Consolidated Drain discharge (Appendix E).

Analysis of post-construction daily temperature fluctuations in Portage Creek indicated that daily instream fluctuations are primarily a result of ambient air temperature change but can be exacerbated by storm events (Section 6.4.2). For example, Consolidated Drain increased the daily fluctuations in Portage Creek up to 2.4°C (September 8, 2001) during wet weather, most likely due to the unmitigated runoff carried by the Schuring Drain. The stormwater treatment ponds and wetland mitigated daily thermal impact of stormwater runoff from the contributing subwatershed by storing runoff and allowing equilibration with air (Section 5.3). During dry weather, vegetative cover played a significant role in regulating daily stream temperature variations by shading the stream channel. However, due to its relatively small baseflow, the Consolidated Drain overall did not have much effect on the daily fluctuations of Portage Creek during dry weather.

Two conclusions are drawn here regarding the role of traditional stormwater treatment ponds in thermal pollution abatement: 1) though ponds with available capacity to capture volumes from single storms may abate acute stormwater thermal shock to receiving streams, those at full capacity or those experiencing back-to-back storms may exacerbate the problem by flushing out thermally enriched stormwater from prior events; and 2) slow-release ponds can be a source of chronic thermal enrichment to the receiving stream (due to solar heating) if discharged directly to surface waters. The flow component and receiving stream temperature regime analysis illustrates the importance of groundwater baseflow in maintaining the thermal regime of the receiving stream. Baseflow mitigates the potential for chronic thermal enrichment associated with a nearly continuous return of treated stormwater from wetlands that achieve ambient air temperature. In addition, riparian shading is crucial in preserving the positive effect of baseflow mitigation.

8.5 Temperature Equivalent

A TE concept was examined in Section 6.3 that combines the temperature and flow volume in defining a parameter that quantifies the thermal pollution contribution of various drainage areas in a watershed. Having units of degrees Celsius, TE is the flow-weighted average temperature of the flow component calculated in a pre-defined time period. TE was applied to stormwater runoff and stream flows of the Consolidated Drain watershed for the August-September (2002) comparative period. Results showed that stormwater runoff from the Crossroads Mall had a TE value of 23.3°C, 5.8°C above Portage Creek TE, indicating the significant source of thermal pollution to the receiving stream from expansive areas of impervious surfaces.

With a prolonged detention period in the sediment forebay and wet detention pond, the TE value of the detained stormwater rose to 24.1°C, an indication of the warming effect of these open ponds. The wetland, in contrast, was able to cool the outflow from the wet pond by 2.5°C TE. Overall, the treatment system lowered the TE of collected stormwater by 1°C, with the wetland cell accounting for all the cooling. However, with regard to the receiving stream, the three-cell treatment system is still considered as a source of thermal pollution as its outflow TE was 4.1°C higher than Portage Creek. This in-system TE pattern confirms the challenges of mitigating heated runoff in open water treatment systems.

The TE value of the Schuring Drain flow was the lowest among all the flow components during the two-month period in 2002, including that of Portage Creek. This means that the Schuring Drain generally constituted a thermal pollution mitigating source to the wetland outflow. This was particularly significant during dry weather conditions when the wetland continued to discharge to the Consolidated Drain. Groundwater baseflow of the lower portions of the Consolidated Drain channel also served to offset the warmer wetland outflow. Immediately above its confluence with Portage Creek, the Consolidated Drain TE value was 17.4°C, slightly lower than the creek. This suggests the post-construction Consolidated Drain watershed did not add thermal pollution to Portage Creek on a flow-weighted thermal load basis over the two-month comparative period in 2002. Overall, TE calculations demonstrate that the Regional Stormwater Quality and Trailways Facility, with the help of groundwater baseflow, can successfully mitigate stormwater thermal loading to Portage Creek during summer months.

TE is easily measured with short interval temperature and flow instrumentation. The value of the TE concept is illustrated by its ability for locating thermal pollution and mitigation sources in a watershed including performance of individual BMPs (e.g., ponds), treatment systems, drains, streams, and groundwater baseflows. Though short-term TE values are not likely as useful given variable event size timing and intensity, long-term applications, as in this study, allow for comparison of thermal characteristics of these system components resulting in a long term temperature index. Ultimately, it is this temperature index that may prove most useful in identifying thermal mitigation opportunities in urban coldwater stream habitats.

8.6 HEATRAN Runoff Temperature Model

A conceptual thermal enrichment model called HEATRAN (Chapter 7.0) was developed for direct runoff temperature simulation in the Consolidated Drain watershed. The model framework was tested for the Crossroads Mall drainage with two fully monitored afternoon storm events. HEATRAN computes runoff temperature directly, using a distributed heat balance for most components in a drainage. The central idea of the model is the combination of heat balance and mass balance. HEATRAN covers principal heat transfer processes such as radiation, evaporation (sensible heat), and convection, and physical drainage system components such as impervious pavement, open channels, closed conduits, confluences, ponds, and wetlands.

All situations or hydraulic elements involved in the drainage are generalized into a single description, and applied to four broadly similar flow situations or elements: 1) overland flow; 2) closed conduit or pipe flow; 3) open channel flow; and 4) open storages (ponds and wetlands). HEATRAN discretizes a watershed into the four categories of hydraulic elements and connects them as a network. Therefore, it is applicable to an infinitely large and complex drainage network system. The energy budget is analyzed for every element in the system.

As a deterministic and continuous temperature model, HEATRAN is reasonably efficient and computes temperature changes in a comprehensive way for a typical urban watershed. Model analysis and sensitivity tests identified several factors influencing runoff temperature including:

- For subcatchment elements, rainfall characteristics (e.g., duration, amount, and temperature) and temperature difference between rainfall (or air) and the ground surface are the two most important controlling factors for the heat transfer processes in wet weather conditions, where the effect of radiation and evaporation is limited.
- For the convective heat transfer in the subcatchment element, rainfall intensity and flow depth increase runoff temperature.
- The pipe network has a cooling (or negative) thermal enrichment effect.
- Apart from the convective heat transfer between runoff and the pipe wall, an extra heat loss exists inside pipes, related to flow depth.
- Several convective heat transfer coefficients are recommended for further development or examination (see Section 7.5).

Based on experiences in the development and testing of HEATRAN, the following implications are conceptually made for the management of thermal enrichment of urban stormwater. At the outset, it should be noted that for impervious subcatchment elements, the sensitivity analysis and the analysis of model results showed that the two dominant parameters for heat transfer during wet periods were rainfall characteristics and the initial temperature difference between rainfall (or the air) and the ground surface. In practice, the only controllable factor is the initial temperature difference between rainfall and the ground surface. As such, it would be desirable to manage heat at the source by increasing the reflective ability of the ground surface and increasing shading. In addition, overland and drain flow can be managed in certain ways to reduce heat transfer to runoff. Specifically, a few suggestions can be made where runoff is subject to overland runoff including the following.

1. Shading can reduce surface heat absorption from insulation, and thus decrease the overall surface temperature. In the model, net radiation in shaded areas can be ignored, assuming that a given percentage of canopy intercepts the same percentage of incoming solar radiation (Weatherbe, 1995).

2. Changing the color of the surface can also alter reflective ability. Lighter color has a higher reflective ability, or lower solar adsorption. Use of lighter-color surface material decreases the overall surface temperature, resulting in lower runoff temperature.

3. It is a sound practice to reduce the total area of impervious black surfaces, i.e., asphalt pavement and roofs. The smaller the impervious black surface area, the less the thermal effect.

4. Pervious surfaces, especially permeable pavement, would be useful for low traffic load areas such as parking lots. Permeable pavement can reduce not only the amount of direct runoff but also runoff temperature. For example, the temperature increase in surface runoff from Eco-Stone was found to be less than that from an asphalt surface (Verspagen, 1995). Although this study did not quantify the difference between pavements in simple terms (as the temperature reduction was highly dependent on how much rain actually infiltrated), it did find that infiltration was substantial and there was no or little heat contribution from properly maintained permeable pavements. If vegetation takes roots in permeable pavement, the cooling effect may be greater because of higher reflectivity, evapotranspiration, and shading. Vegetation also increases surface cooling.

5. Depth and velocity of overland flow are direct responses to rainfall and affect heat transfer processes. Because of the likelihood of ponding and associated slower evaporation, flat slopes and the resulting increased flow depth will increase runoff temperature. Ponded areas on the pavement could result in warmer runoff.

For pipe elements in this research, HDPE was found to have a lower convective heat transfer coefficient than concrete, probably due to a lower thermal conductivity. Underground pipe is usually in a cooler environment than open channels. Runoff from overland flow will lose heat inside the pipe. Therefore, a higher convective heat transfer coefficient will help to transfer more heat from the runoff to the pipe wall, implying that concrete pipe might be a better choice for thermal enrichment management. Similar remarks apply to water flowing through a barrier of rocks, or through any arrangement of covered flow paths with a favorable arrangement of contact surfaces.

8.7 Aquatic Community Assessment

The goal of urban stormwater treatment ultimately is to restore or protect healthy, natural aquatic biota in urban streams. Suitable thermal regime is one of the most important conditions that aquatic biological communities require for sustainability in coldwater streams. Because thermal regime change is only one of many ecological stresses leading to community structural shift, it is difficult to isolate thermal impacts on the biota from multiple stressors (e.g., nutrients and sediment).

In this study, fish, benthic macroinvertebrates, and periphyton were sampled at multiple sites during the four-year period (Appendix C). A newly developed multi-species stream classification system (Wehrly et al., 2003), based on regionally specific relationships between stream thermal regimes and fish community structure in the Lower Peninsula of Michigan, was applied to the fish data analysis as related to the impact of the stormwater treatment system. By identifying mean July temperature categories (cold, <19°C; cool, 19-22°C; and warm, ≥22°C), and July temperature fluctuation categories (stable, <5 °C; moderate, 5-10°C; and extreme, ≥10°C) a 3 × 3 matrix was developed with nine categories. Stream thermal habitat preference and suitability ratings were developed for common Michigan stream fish and separated into coldwater, cool water, and warm water guilds.

Based on its thermal regime and this new fish community classification system, Portage Creek was considered coolwater-moderate at sites both upstream and downstream of the Consolidated Drain in both the pre- and post-construction periods. The upstream assemblage was dominated by coldwater guild species, species occupying similar ecological positions within the same habitats. The downstream assemblage was more balanced with coldwater and cool water guild species. Furthermore, the post-construction downstream assemblage more closely resembles the upstream with a shift toward greater dominance in coldwater guild species and lesser contribution of cool water guild species. This suggests improvement in the downstream thermal regime associated with the 2001 completion of the treatment system. However, because most coldwater guild species are also pollution sensitive, long-term sampling is necessary to establish any conclusive relationship capable of separating the effects of thermal regime improvements over other common pollutants associated with the treatment system operation.

Changes in macroinvertebrate and periphyton metrics measured for pre- and post-construction Portage Creek sites upstream and downstream of the Consolidated Drain were attributed to natural annual variation based on consistent variation across stream sites. Therefore, data did not show obvious changes in Portage Creek associated with the construction of the Consolidated Drain stormwater treatment facility. Nevertheless, the Consolidated Drain channel developed a healthier post-construction macroinvertebrate community with the appearance of caddisfly species and a better community balance. In addition, periphyton community dominance shifted from warm water (blue-greens) to coldwater (diatoms) species after construction.

8.8 Thermal Enrichment and Stormwater BMPs

This study established that effective infiltration and adequate shading are the two key design criteria for successful protection of coldwater streams from stormwater thermal pollution. Analysis of temperature regime changes and daily temperature fluctuations in the coldwater receiving stream suggested that despite capture of the first flush of stormwater runoff, stormwater BMPs, such as open water detention ponds, may exacerbate acute thermal impacts of summer storms. The slow release of thermally enriched water from these ponds renders them potential chronic thermal pollution sources. This is particularly true for watersheds where groundwater-fed baseflow is the primary source of cool water (temperature lower than the air) and summer solar radiation is intense.

Well-vegetated wetlands and infiltration basins (or other similar BMPs) are important elements that can achieve conventional treatment goals (i.e., pollutant removal and hydraulic mitigation) without introducing high temperature outflow to coldwater receiving streams. Researchers have observed, however, that moderately developed watersheds with BMP ponds and stormwater basins have less aquatic biodiversity than watersheds with similar levels of development and no BMPs (Way, 2003). Although no direct cause-effect relationship was found, it was suspected that ponds may have been culpable for this diversity loss. This study provides quantitative evidence to at least partially support this speculation.

One may argue that well-shaded open detention ponds can prevent thermal enrichment of stored stormwater. Although this study showed that vegetation shading can provide thermal mitigation in vegetated wetlands and open conveyance channels, it will have only limited effect in open ponds for the following two reasons. First, for practical reasons, it is difficult to provide complete shade coverage for open ponds, particularly large ponds in regional treatment systems. Vegetation planted on the perimeter of a large pond provides shading to only a small portion of the pond. This shading is most limited during afternoon hours of summer days when the sun is high and heating of ponded water is greatest. Second, it takes time for vegetation to mature. To shade a pond, mature trees with extensive canopies are necessary, which means years may pass before any substantial shading can be established.

From the perspective of source control, shade trees in impervious areas in urban centers can result in lower pavement temperature as evidenced by monitoring (Chapter 4). In addition, trees or other vegetative shading can be used to mitigate stormwater thermal impacts in open conveyance channels. TE analysis demonstrated, however, that open ponds, as a primary source of stormwater thermal enrichment, will not provide sufficient relief to stormwater thermal enrichment despite shading.

8.9 Framework for Mitigating Stormwater Thermal Enrichment

With the support of data analysis and the HEATRAN model, a design framework is conceptualized here to address stormwater thermal enrichment in urbanizing watersheds with coldwater receiving streams. In principle, the primary goal of heat load reduction is to reduce surface runoff volume, both before and after runoff reaches stormwater treatment facilities. Infiltration should be the focal point of thermal pollution mitigation in runoff abatement and runoff treatment. Such an approach is compatible with the traditional goals of most stormwater treatment efforts for conventional pollutants and hydraulic mitigation. In coldwater streams where groundwater-fed baseflow is what maintains low water temperature of the stream during warm months of the year, runoff volume minimization is particularly important. After runoff volume reduction efforts are maximized, other options can be examined to reduce runoff temperatures. These secondary options will involve riparian corridor and BMP shading and to a lesser extent, reducing the heat stored in impervious building materials. Where available, existing baseflow sources (surface or subsurface) can be utilized to offset treated runoff that has reached ambient air temperatures and is being discharged back to a receiving water via moderated surface flows.

Specifically, the conceptual framework can be outlined as follows according to the flow pathway of surface runoff in an urbanized watershed:

Before surface runoff reaches conveyances, treatment facilities, or streams:

- Restore natural flow regime:

  1) Decrease impervious surface area to reduce overland flow
  2) Promote onsite infiltration and re-use of treated stormwater
  3) Increase interception of rainfall by vegetation

- Reduce surface temperature:

  1) Choose light-colored building materials with highly reflective surfaces to reflect solar radiation and reduce heat sorption
  2) Provide shading to impervious areas to reduce heat sorption

In conveyances, treatment facilities, or streams:

- Promote infiltration:

  1) Build conveyances and treatment facilities that promote infiltration of treated stormwater
  2) If possible, use infiltration basins; otherwise, choose retention over detention ponds

- Promote temperature loss in underground pipe networks:

  1) Choose pipe material with high thermal conductivity (e.g., concrete) over that with lower thermal conductivity (e.g., HDPE) to promote heat loss in underground conveyance
  2) Reduce depth and contact time of runoff with surfaces and minimize ponding on impervious surfaces
  3) Consider deep discharges from ponds over surface water discharges if suitable from a hydraulic and pollutant removal perspective

- Reduce direct solar inputs:

  1) Maximize shading of ponded water surfaces by including well-vegetated wetlands with open wet ponds
  2) Reduce exposed surface area of channels and outlet structures
  3) Promote the growth of riparian and channel-bed vegetation
  4) Align ponds in a north-south orientation to minimize solar radiation

- Utilize existing groundwater baseflow:

  1) Construct off-line treatment cells and mix treatment outflow with groundwater baseflow
  2) Separate groundwater baseflows from treatment systems

8.10 Future Studies

Four important aspects of the stormwater thermal issue that are not fully explored in this study are: 1) the effects of temperature on the chemical and biological processes in stormwater ponds and wetlands that are vital in pollutant removal; 2) the non-summer thermal regime changes in receiving streams due to stormwater ponds; 3) the effects of thermally enriched stormwater infiltration on groundwater temperature regimes; and 4) the applicability of relationships developed in this study to other geographic settings.

Temperature is a critical factor in chemical and biological reactions. Kinetically, most chemical reactions take place at a higher rate when temperature increases, which can improve the pollutant removal efficiency for pollutants such as nitrogen. Kadlec and Reddy (2001), for example, reported that a 10ºC increase in temperature can result in up to four-fold increase of nitrogen removal in treatment wetlands. Results from this study showed that temperature increases of such magnitude can happen in open stormwater ponds (e.g., Figures 5-2 and 5-4).

In terms of non-summer thermal impacts, even fewer studies are available to give a clear direction on research needs. Biologically, heat stress to coldwater fishes is most severe in the warmest months of the year when fish are most susceptible. This includes the physical stream conditions of low flow and high temperature. During periods of ice cover in northern climates, such as those in Michigan, water temperatures in wetlands (and likely ponds) are slightly more than 0ºC (Kadlec and Reddy, 2001). Data from this study indicate that groundwater feeding into the stream system in the treatment area has a stable temperature of about 15°C year-round. Occasional snow melt during the winter and the annual spring snow melt generate runoff much cooler than this groundwater temperature. As this study collected non-summer temperature data throughout the treatment system, a natural next step for a continued stormwater runoff thermal study would be to examine the effect of runoff on the receiving stream during winter months.

Groundwater temperature regimes are likely influenced by the infiltration of thermally enriched urban stormwater. These effects are likely complicated by influences at larger scales such as the urban heat island effect. Long-term monitoring of groundwater temperature should be considered in future efforts.

Empirical derivations from this study should be tested in other settings and potentially expanded to other watersheds. The statistically significant relationship between reduction in stormwater volume and reduction in heat loading could serve as a useful guide for stormwater treatment design if it can be successfully extrapolated to other systems. Additional research should focus on spatial (e.g., elevation, latitude, land use) or temporal (length of season) drivers in this relationship. If verified in other settings, the concept of TE might aid in the design of comprehensive thermal mitigation strategies for individual BMPs, treatment systems and watersheds where thermal enrichment is a concern.

Because watersheds vary in their natural characteristics and state of physical development, it would be instructive to also test the mechanistic modeling observations derived from this research in other settings. Validated model assumptions could be used to derive site-specific planning standards for landscaping (shading) as well as design of hard surfaces considering materials and finished grade. A validated mechanistic framework could also be coupled with existing stormwater flow models for large-scale urban applications where more sophisticated efforts are in place that utilize such tools. This could be particularly useful in designing appropriate systems in urbanizing areas near coldwater habitats. It could also have applicability in areas with existing infrastructure that need to implement stormwater controls to address TMDLs, Phase II Stormwater Regulations, or other environmental restrictions directly or indirectly associated with temperature.

In summary, efforts in this study have identified a range of needs and opportunities that can be addressed in future research. Temperature, as an easily measured parameter, should be considered as an integral component of any stormwater monitoring program where thermal enrichment may be problematic for receiving streams. It is apparent that efforts to mitigate thermal enrichment have benefited the coldwater receiving stream in this research setting. Inexpensive temperature monitors placed in conjunction with other more traditional monitoring devices for sample collection and flow measurements have confirmed the obvious as well as the subtle dynamics of thermal enrichment throughout the drainage area and stormwater treatment system components. This information is critical for designing effective stormwater controls and managing conditions that impact the thermal regime of coldwater receiving streams.

APPENDIX A

Mark S. Kieser
Nicole Ott

MONITORING OF THE CONSOLIDATED DRAIN AND THE CITY OF PORTAGE STORMWATER QUALITY AND TRAILWAYS FACILITY

LIST OF TABLES

LIST OF FIGURES

A.1 Background

Prior to July of 2001, the Consolidated Drain No. 1 was an open channel waterway that drained approximately 463 acres in the heart of the City of Portage's commercial center (Portage Commerce Square) (Figure A-1). The drain was constructed by the Kalamazoo County Drain Commissioner in 1963 and was turned over to the City of Portage in 1972. The watershed of the drain has since become highly urbanized. The Consolidated Drain discharges to Portage Creek in the vicinity of the Celery Flats Interpretive Center at Garden Lane (Figure A-2).

Stormwater loads from this drain had been identified as the largest single contributor of pollutants to Portage Creek within the City of Portage. The Consolidated Drain (and a contiguous urban drainage channel, the Schuring Drain) discharge to a section of Portage Creek designated as a coldwater habitat where trout have historically been stocked by the State of Michigan. This area of the creek also serves as a major focal point of the City's Bicentennial Park and Celery Flats Interpretive Center, as well as a primary trailhead for a non-motorized public trailway. Preserving the natural resources of Portage Creek is an important objective for the quality of life in Portage.

Stormwater flows within the Consolidated Drain were observed to be very flashy (prone to sporadic, unstable discharges, including sudden, high flows of stormwater), causing erosion and habitat disruption in Portage Creek. Additionally, during summer storm events, stormwater became heated as it flowed over warm impervious surfaces. This introduction of warm water to a coldwater habitat was potentially harmful to the trout population, along with other stream biota. Table A-1 illustrates the originally estimated annual pollutant loads reaching Portage Creek from the drain (K&A, 1999).

Table A-1. Consolidated Drain estimated annual pollutant loads (pounds) to Portage Creek. (Based on published loading estimation techniques, Wayne County, 1998.).

Total Suspended Solids	Total Phosphorus	Kjeldhal-N	Nitrate+ Nitrite-N	Lead	Copper	Zinc	Cadmium
243,313	694	3,723	1,802	335	101	738	5

A.1.1 Regional Stormwater Quality and Trailways Facility

Based on a stormwater treatment feasibility study completed by KIESER & ASSOCIATES (K&A) in 1999, a regional detention facility was conceived to decrease pollutant loading to Portage Creek, mitigate hydraulic flashiness and reduce flooding. Stormwater from a very small fraction (~3%) of the 463-acre drainage watershed was previously detained by a limited number of small or failing detention basins. These basins tended to release detained stormwater at similar times, still contributing to flashy stormwater loads to Portage Creek delivered from the remainder of the drainage area. By eliminating these basins and conveying stormwater from the watershed to one regional treatment facility, the stormwater could potentially be treated and detained over an appropriate holding time.

As a part of the Portage Commerce Square Enhancement Initiative, the "Regional

Stormwater Quality & Trailways Facility" was constructed in 2000 and 2001 to treat off-line, a two-year, 24-hour storm event for the entire watershed in a centralized location. First flush (up to one inch of rainfall runoff) volumes are captured and treated in this system. Flashiness is reduced as the water travels through the new treatment system and is slowly released to Portage Creek via a reconstructed Consolidated Drain flood channel. The outlet of the treatment system releases treated stormwater to Portage Creek in a steady flow rate.

In the northern portion of the new treatment facility, a 42" diameter pipe transports stormwater runoff from the Consolidated Drain Watershed to the sediment forebay (Figure A-3). During heavy rains (greater than a two-year, 24-hour event) stormwater overflow occurs from two concrete weir structures directly into the reconstructed, vegetated surface drain channel. Downstream of the weir structures, a reflecting pond and fountain contribute a one cubic foot per second (cfs) baseflow to the reconstructed downstream channel during summer months. During the winter months the reflecting pond serves as a refrigerated ice rink (with no baseflow).

The stormwater conveyed to the 2.1-acre sediment forebay is first treated by removal of floatable materials and coarse solids. Stormwater is then detained in a 12.0-acre wet detention pond for a minimum of 11 hours (average of 72 hours) for removal of fine particles. Much of the existing vegetation in the wet pond remained undisturbed during the construction phase of the project. Existing surface muck soils were stockpiled and re-used on new berms which allowed the original native seed bank to revegetate. Berms were additionally planted with native wetland and upland species.

The discharge from the wet detention pond is conveyed to a 11.6-acre wetland for final polishing over a duration of 5 to 10 days. This wetland complex existed prior to construction of this new facility. Since the treatment system was brought on-line in August of 2001, this wetland area has flourished with periodic inundation of up to a maximum 1' depth. Outflow from the wetland is aerated over a shaded limestone rock outfall as it discharges into the reconstructed drain channel/floodway. This new channel also conveys a one cfs baseflow from the upstream reflecting pond and meanders through the new treatment facility. Log deflectors and rock features were added to stabilize the banks and provide habitat diversity. The Schuring Drain, a local tributary channel which conveys runoff from a 351-acre watershed (consisting primarily of greenhouses and residential areas to the southwest of the Consolidated Drain Watershed) was also reconstructed and is now discharged (without treatment) into the Consolidated Drain upstream of the Wetland Outlet. Figure A-3 illustrates the configuration of the newly constructed system with sampling locations included.

A.2 Water Quality Monitoring

A.2.1 Sampling Summary

Preconstruction sampling was conducted from November 1999 to October 2000 at two locations along the Consolidated Drain in order to determine loading to Portage Creek (Figure A-2). ISCO automated samplers were placed at locations both downstream and upstream of the Schuring Drain confluence with the Consolidated Drain. The Upstream automated sampler was removed in late 2000 with construction of the new channel. The Downstream automated sampler, located along Garden Lane, remains on-site today. In addition to pre-construction sampling, it was utilized to collect samples from three "during-construction" storm events in 2001 and seven post-construction storm events in the summer of 2002. During the post-construction sampling, two additional automated samplers were installed at the Sediment Forebay Inlet structure and at the Wetland Outlet structure, to ascertain the load reduction capabilities of the treatment system. Grab samples were also collected at locations throughout the treatment system and in the Schuring Drain in order to further understand each component of the system. (See Figure A-3.)

A.2.2 Sampling Methods

A Quality Assurance Project Plan (QAPP) was approved by the Michigan Department of Environmental Quality (MDEQ) in May of 2000. Sampling activities associated with Section 319 funding have generally followed the procedures laid out in that plan. Monthly internal audits were completed to assure that field equipment was properly maintained. Performance audits were also completed upon receipt of analytical data, to check the results returned by the analytical laboratory. Results of the audits have been included in the quarterly reports, submitted to MDEQ throughout the duration of the project.

During the pre-construction phase, the samplers were used to collect water samples from the drain during rain events. The sampler at the Upstream location was programmed to begin sample collection when either the water level (as measured with an ISCO 720 Submerged Probe Module) in the drain increased approximately ½" above the recorded level prior to the event or when precipitation (as measured with a 674 ISCO Rain Gage) amounted to at least two-tenths of an inch during a 30-minute time period. The Downstream automated sampler began sampling only when the water level was at least ½" above the value preceding the event, since no rain gage was connected to the Downstream automated sampler.

Once the instruments were enabled by the specified parameter(s), the pump automatically purged the suction line with air and rinsed it with the water from the drain three times. Once sample collection began, a sample was collected every one and one-half hours until all 24 bottles were filled. Filling of each one-liter bottle took approximately 20-30 seconds. In anticipation of a rain event, ice packs were placed in the center of the ring of the bottles to ensure the water samples remained at or below the drain water temperature. Two one-liter bottles were filled for each discrete grab sample. This allowed for collection of enough sample volume to analyze each sample for the MDEQ's MA, GA and GN analytes (Table A-2).

Table A-2. Analytes sampled at the Consolidated Drain Treatment Facility, grouped by MDEQ Code.

MA Analytes	GA Analytes	GN Analytes
Arsenic	Total Phosphorus (TP)	Total Suspended Solids (TSS)
Barium	Nitrate + Nitrite-N	Soluble Reactive Phosphorus (SRP)
Cadmium	Ammonia-N	Nitrite-N
Chromium	Total Kjeldahl-N (TKN)	Chloride
Copper	-	-
Lead	-	-
Mercury	-	-
Selenium	-	-
Silver	-	-
Zinc	-	-
Hardness	-	-
Sodium	-	-

During periodic visits to the sampling locations, flows were measured using a direct discharge cross-section approach with a Marsh-McBirney Flow-Mate Model 2000 portable velocity meter. The depth of the water and the width of the drain channel at the water line were measured at the opening of the downstream side of two existing culverts. Velocities were measured in triplicate at approximately two-thirds of the water depth across five horizontal stations. The cross-sectional areas of each horizontal segment and the average velocity were multiplied and summed for each segment to calculate a discharge value for each cross-section. The calculated flows and measured water depths were then utilized to generate stage-discharge curves for each station (Attachment A-1). Two stage-discharge curves were established at the Downstream location to represent the pre-construction and post-construction sampling periods.

Flow from the Wetland Outlet of the constructed treatment system now passes through a weir structure before entering the Consolidated Drain. A level probe was installed at the crest of the weir in the Spring of 2002, which recorded levels at five-minute intervals. Grab samples were collected by field personnel during the spring and early summer from the Wetland Outlet. An automated sampler was installed at this location in July 2002 which also recorded level measurements at five-minute intervals. Flow was calculated using recorded water depths and the engineering equation for a broad-crested, square notch weir.

An automated sampler was also installed at the Sediment Forebay Inlet structure in late July 2002. An ISCO 4150 area-velocity meter was installed at the end of the inlet pipe and recorded both level and velocity at five-minute intervals. Flows were calculated internally by this equipment based on the recorded depth, velocity and known pipe diameter. The automated sampler was programmed to begin sampling when flow exceeded two cfs. This station responded rapidly to rainfall (within minutes). Flows were visually observed to peak quickly in response to the beginning of an event and dissipate shortly after rainfall stopped corresponding to the highly impervious, storm sewered drainage area.

The ISCO 6700 Portable Samplers were visited before and after each sampled rain event. After all desired sample bottles had been collected, the samplers remained disabled and only recorded date, time and water level at each site (also velocity at the Sediment Forebay Inlet). Prior to a sampling event, the samplers were reset and the automated sampling routine was checked to ensure that the proper sampling schedule was programmed. The water levels at the sampling locations were measured manually, and each instrument was corrected as needed to ensure instrument readings and manual measurements coincided.

Most samples collected for laboratory analysis were processed in accordance with the MDEQ laboratory protocols. When the MDEQ laboratory could not accept samples, they were sent to TriMatrix Laboratories (the MDEQ's overflow laboratory) in Grand Rapids, Michigan. The November 1999 and March 2000 preconstruction events were sampled by KAR Laboratories in Kalamazoo, Michigan and by Upstate Freshwater Institute (UFI) in Syracuse, New York. Duplicate analyses were performed on every tenth sample for each parameter requested from a sampling event. Samples were analyzed for the ten Michigan heavy metals (see first ten analytes in "MA Analytes" column of Table A-2.), ammonia, total Kjeldhal nitrogen (TKN), nitrate + nitrite, nitrite, Total Phosphorus (TP), Soluble Reactive Phosphorus (SRP), sodium, chloride, hardness, and total suspended solids (TSS).

A.2.3 Quality Assurance/Quality Control

It was noted that mercury was non-detect in all sampling data returned by the analytical laboratory. Therefore, it was dropped from the suite of parameters analyzed in the summer of 2002. It was also noted that the TP results returned for the Wetland Outlet were greater than expected. Therefore, several split samples were sent to UFI. TP comparisons between UFI and TriMatrix Laboratories and between UFI and the MDEQ Laboratory are presented in Attachment A-2. Generally speaking, these results do not suggest any statistical differences between laboratory analyses.

Samples collected during two events in August 2002 were sent to TriMatrix Laboratories, MDEQ's overflow laboratory. TriMatrix Laboratories used an elevated detection limit for TSS, ranging from 12- 25 mg/L. Therefore, several Wetland Outlet data were reported as less than 25 mg/L. Analytical data from July 2002 suggest that the TSS concentrations at the Wetland Outlet were typically below 4 mg/L. Therefore, a concentration value of 25 mg/L did not accurately describe the outflow. Using 25 mg/L in the loading calculations would produce an overestimation of the TSS load from the Wetland Outlet. Therefore, TSS loading could only be accurately calculated for eight of the ten defined events at the Wetland Outlet.

A.3 Data Analysis

A.3.1 Preconstruction Data

Fourteen events were analyzed at the Upstream location during pre-construction sampling from November 1999 to October 2000. Sampling at the Downstream location was extended into the "during-construction" phase in 2001 and included 19 events. Concentration values returned by the analytical laboratory were recorded in separate spreadsheets for each event. Water levels, recorded at five-minute intervals, were converted into flows using the stage-discharge relationships developed for each sampling location (Attachment A-1). Concentration values were matched with flow values at each corresponding sampling time. Five-minute concentration values were then interpolated between each measured time point. The measured and interpolated concentrations were multiplied by the corresponding flow value, the appropriate time step (generally 300 seconds) and a conversion factor in order to calculate a pollutant load in pounds for each parameter over a five-minute period. The five-minute loads were then summed over the duration of the sampled event. Attachment A-3 presents the pollutant load summaries for each preconstruction sampling event.

A rain gauge was located at the Upstream automated sampler during the pre-construction phase. Rain gauges were also located in various retention basins throughout the City of Portage and in the wetland treatment area (during the post-construction phase). A measured rainfall depth was recorded for each sampling event. The analyte load for each event was regressed by both a linear and power function against the rainfall depth in order to determine if a significant relationship between pollutant loading and rainfall existed (Attachment A-3). The relationship having the greatest R^2 value (between the linear and power equation for each analyte) at each sampling location was then utilized to compute an annual pollutant load. These estimates are based on the observed rainfall frequencies for various sized events in Kalamazoo, Michigan (Table A-3).

Table A-3. Rainfall frequency distribution for Kalamazoo County.

Average rainfall per 24-hr event (inches)	Frequency of occurrence
0.25	25.2
0.50	13.8
0.75	8.5
1.00	4.7
1.25	3.5
1.50	1.0
1.75	1.1
2.00	1.0
2.25	1.1
2.50	0.7
2.75	0.3
3.00	0.2
3.25	0.1

Notes: 100-year dataset collected at the Kalamazoo Psychiatric Hospital, Kalamazoo, MI.

Annual loads of those parameters estimated in Table A-1 (TSS, TP, TKN, nitrate+nitrite-N, lead, copper, zinc and cadmium) were calculated for the Upstream location during the preconstruction phase and for the Downstream locations during the pre- and during- construction phases. The annual pollutant loads from the Schuring Drain were calculated by subtracting the Upstream load from the Downstream load. Event mean concentrations (EMCs) for the Upstream and Downstream locations were calculated by dividing the total mass load for each sampled event by the total flow volume for each event. Table A-4 presents the estimated annual loads for each parameter. Tables A-5 and A-6 present the EMCs for each station.

The flashiness of the Downstream location was also evaluated to compare to the post-construction hydraulic behavior in the Consolidated Drain. Flashiness was calculated for the summer sampling period, June to September 2000, by computing the average level and the amount of time that the hydrograph was above that average level. An undisturbed channel is expected to have levels (and, hence, flows) above the average approximately fifty percent of the time. A flashy channel experiences rapid rises and sharp declines in flow, resulting in a smaller percentage of time when levels are above the average.

Table A-4. Annual wet weather pollutant loading estimates (pounds) based on measured data. See Figures A-2 and A-3 for station locations. Station identification numbers in parentheses refer to monitoring locations in Figure B-1.

Station	TP	TSS	TKN	Nitrate+ Nitrite-N	Cadmium	Copper	Lead	Zinc	Number of Events
Upstream preconstruction (approximately at 27)	114	21,351	786	809	not detected (nd)	4.5	5.5	33.3	14 (tp/tss) 8 (remaining analytes)
Downstream pre- and during-construction (3)	219	39,283	980	1,092	nd	6.6	9.3	41.8	13(tp/tss)* 9 (nitrogen) 10 (metals)
Schuring Preconstruction (23)	105	17,932	194	283	nd	2.1	3.8	8.5	downstream minus upstream
Wetland Outlet (19)	46.4 (July 29-Sept. 9)	30,341	2,380	137	nd	2.3	nd	nd	10 8 (TSS)**
Sediment Forebay Inlet (28)	301 32.4 (July 29-Sept. 9)	110,956	1,229	817	0.5	17.8	34.8	123.4	7***
Downstream post-construction (3)	150	94,419	718	369	nd	5.3	7.1	26.4	6****

Notes:

*Fourteen events were sampled and utilized for the EMC calculation. One outlier was removed, leaving thirteen events for the annual loading calculation.

**Ten events were defined at the Wetland Outlet. However, elevated TSS detection levels were reported as non-detect by the laboratory. Therefore, only eight events were used to calculated the TSS load.

***Nine events were sampled. However, two sets of back to back events occurred. Only the first event of each set was used in the load vs. rainfall calculation, allowing seven events to be utilized to calculate an annual load.

****Seven events were sampled. However, one was based on snowmelt and rainfall and could not be used for the load vs. rainfall regression.

Table A-5. Average and maximum observed TP and TSS EMCs for each sampling station.

Station	Average TP (mg/L)	Maximum TP (mg/L)	Average TSS (mg/L)	Maximum TSS (mg/L)	Number of Events
Upstream preconstruction	0.15	0.38	24.4	230	14
Downstream pre- and during-construction	0.18	0.67	32.9	1,140	14
Upstream* post-construction	0.04	0.13	17	59	(12 samples)
Schuring post-construction	0.19	1.14	115.5	1,010	(12 samples)
Sediment Forebay Inlet	0.16	1.48	52.6	220	9
Wetpond Inlet	0.11	0.19	19.8	58	(4 samples)
Wetland Inlet	0.20	0.27	14.0	21	(6 samples)
Wetland Outlet	0.34 (summer) 0.12 (feb-may)	0.59 (summer) 0.22 (feb-may)	9.8 (summer) non detect (feb-may)	21 (summer) 9(feb-may)	(8 samples, feb-may) (116 samples, summer TP) 8 events, summer TSS**
Downstream post-construction	0.26, using load/volume method (0.16, geometric mean) (0.03, dry weather)	0.77	173, using load/volume method (77.9, geometric mean) (non detect, dry weather)	890	7

Notes:

*Consolidated Drain Upstream grab samples were collected immediately above the confluence with the reconstructed Schuring Drain. These would represent baseflow plus weir overflows.

**Due to elevated TSS detection limits used by TriMatrix Laboratories (MDEQ overflow laboratory), only eight events accurately represent low level wetland outflow concentrations.

Table A-6. Nitrogen and metal EMCs for each sampling station.

Station	Average TKN (mg/L)	Average Nitrate + Nitrite-N (mg/L)	Average Cadmium (ug/L)	Average Copper (ug/L)	Average Lead (ug/L)	Average Zinc (ug/L)
Upstream preconstruction	0.64	0.72	<0.2	3.5	4.0	26.0
Downstream pre- and during-construction	0.74	0.84	<0.2	4.9	9.0	30.2
Upstream post-construction*	0.70	0.70	<0.2	13.3	3.5	21.9
Schuring post-construction	1.22	1.62**	<0.2	7.4	7.0	30.1
Sediment Forebay Inlet	0.77	0.37	0.3	9.1	15.6	71.4
Wetpond Inlet	0.80	0.26	<0.2	6.1	8.6	37.5
Wetland Inlet	1.33	0.17	<0.2	4.3	2.6	14.8
Wetland Outlet	1.37	0.04	<0.2	1.1	<1.0	<10
Downstream post-construction	1.1	0.85	<0.2	11.0	13.7	54.5

Notes: *Consolidated Drain Upstream grab samples were collected immediately above the confluence with the reconstructed Schuring Drain. These would represent baseflow plus weir overflows.
**Highest concentrations of nitrate+nitrite-N in the Schuring Drain were collected during dry weather. They were included in the EMC calculation; n=14.

A.3.2 Post-Construction Data

Seven events were sampled at the Downstream location during the post-construction phase, from March to September 2002 (Attachment A-4). Flashiness was also evaluated at the Downstream location and compared to the preconstruction hydraulic behavior. Nine events were analyzed at the Sediment Forebay Inlet from July 29 to September 20, 2002. (See the rainfall hyetograph, Figure A-4.) Flow at the Sediment Forebay Inlet was very flashy, allowing events to be easily defined by a sharp rise and descent in flow rate. (See Sediment Forebay Inlet hydrograph in Attachment A-5.) Event loads, annual loading estimates and EMCs were calculated in the same manner described above for the preconstruction data analysis.

Because the system is designed to detain stormwater and slowly release it to the Consolidated Drain, flow at the Wetland Outlet does not respond quickly to storm events as does the Sediment Forebay Inlet. Often the Wetland Outlet was observed releasing water from a previous rain event while another event began. (See the Wetland Outlet hydrograph in Attachment A-5.) However, events were defined at the Wetland Outlet sampling location by observing flows as they entered the Sediment Forebay Inlet during a given storm. A rise or spike in the Wetland Outflow was used to mark the beginning of an event within the wetland. (The Outlet flow does respond slightly to Sediment Forebay inflow as flow out of the wetpond subsequently pushes stored water out of the wetland.) An event was considered to have ended when the volume of water that entered the Sediment Forebay Inlet had left the system through a combination of evapo-transpiration, limited infiltration and measured Wetland Outlet flow. (A pyranometer located in the wetland recorded solar radiance at ten-minute time intervals. These data were used to measure evaporative losses.) The inputs of rainfall on the surface of the wetland were also factored into this volume calculation. By defining these criteria, events monitored at the Sediment Forebay Inlet were also defined at the Wetland Outlet. In this respect, an event load at the Wetland Outlet was compared to that at the Sediment Forebay Inlet in order to assess the overall pollutant uptake by the system.

Because the wetland slowly releases treated stormwater to the Consolidated Drain, the Wetland Outlet was observed to flow continuously from July 29 to September 9, 2002 (Attachment A-5). Intensive sampling was conducted at the Wetland Outlet location during this time period allowing a load for the entire period (Table A-4) and loads for events occurring during that period to be calculated. Eleven storm events occurred during this period of continual Wetland Outlet flow, six of which were sampled at the Sediment Forebay Inlet, and seven of which were defined for the Wetland Outlet. (The events sampled at the Sediment Forebay Inlet and defined for the Wetland Outlet are labeled on Figure A-4, rainfall hyetograph. The time periods of the events defined at the Wetland Outlet are indicated on the Wetland Outlet hydrograph for the July 29 to September 9, 2002 period in Attachment A-5.) No flows were recorded (i.e., no loading to the Consolidated Drain) directly before and after this time period. Sampling continued at the Wetland Outlet in late September, and three additional events were defined, as outflow resumed. These ten defined events were used to calculate event loads (Attachment A-5), annual pollutant loads (Table A-4) and EMCs (Tables A-5 and A-6), for all parameters except TP, in the same manner as described for the pre-construction events.

TP event loads were not utilized to calculate a TP annual load because the concentrations measured during this time period were not representative of seasonal changes throughout the year. Table A-5 above, provides a comparison of the summer (July to September) EMCs for TP and TSS to the Spring (February to May) EMCs at the Wetland Outlet. The spring concentration values are less than one-half of the summer concentration values for TP. Therefore a TP load for the July 29 to September 9, 2002 period was calculated, as the outflow was intensively sampled. This load (46 pounds) is presented in Table A-4. A Sediment Forebay Inlet TP load was calculated for the same time period. Six storm events, of the 11 that occurred during that time, were sampled. Additionally, three storm events were sampled in late September. The loads from those nine events were used to compute a load vs. rainfall relationship for the Sediment Forebay Inlet (Attachment A-5). That equation was used to calculate the TP loads for the five unsampled events occurring from July 29 to September 9, 2002. The sum of the TP loads from those six sampled events and the five estimated events was 32 pounds, smaller than that for the Wetland Outlet over the same time period (46 pounds). The calculated TP mass load from July 29 to September 9, 2002 was divided by the total Outlet flow volume in order to compute a TP EMC. It was noted that the TP levels were elevated during this time most likely, due to anaerobic conditions within the wetland system. Therefore, these reported concentrations are considered "summer" levels.

Periodic grab samples were also collected by field personnel during the post-construction phase from the Wetpond Inlet, the Wetland Inlet, the Consolidated Drain Channel (Upstream of the Schuring Drain and Wetland Outlet discharges) and from the Schuring Drain Channel. Since automated samplers were not located at these intermediate locations, event loads could not be computed. As a result, the grab samples were utilized to calculate average wet weather concentrations for these locations (also presented in Tables A-5 and A-6).

A.4 Discussion

Annual loading and flashiness at the Downstream location was compared between the preconstruction and post- and during-construction time periods in order to assess stormwater impacts on Portage Creek. Pollutant loads and flashiness were also compared between the Sediment Forebay Inlet and the Wetland Outlet in order to assess the overall pollutant removal by the system. Annual loads were calculated by inserting a range of observed rainfall depths (0.25-3.25") into the appropriate load vs. rainfall depth relationship. These results were then multiplied by the expected frequency of those events (Table A-3). However, since none of the post-construction events sampled were larger than 1.13 inches, the relationship may not hold for rainfall events which exceeded this value.

Annual loads to Portage Creek (measured at the post-construction Downstream location) have been reduced from the pre- and during-construction period for all parameters except TSS. Table A-4 presents the annual load estimates for all stations. This increase in TSS loading is presumed to be associated with the loading from the reconstructed Schuring Drain, which produced a single TSS concentration as high as 1,010 mg/L on May 8, 2002. The August 2, 2002 event at the Downstream station was also noted to have a particularly high TSS concentration (890 mg/L). This value skewed the entire TSS vs. rainfall relationship. The impact of this reading can also be observed in Table A-5. EMCs for the Downstream location were calculated by two methods: 1) the sum of calculated mass loads were divided by the total flow volume for each event, and; 2) an average was simply calculated from the reported concentration values. Method 1 allows the timing of the measured concentration to impact the resultant load. With the Downstream event, for example, the high TSS concentration (890 mg/L) has a significant impact on the EMC. This elevated TSS concentration (890 mg/L) is believed to be accurate, as the turbidity of the Schuring Drain during that event was observed to be very high. The high turbidity observed is believed to be associated with this watershed runoff to the drain. Tables A-5 and A-6 indicate that the Schuring Drain station produced the greatest TSS and nitrate+nitrite-N EMCs.

Flashiness was observed to decrease at the Consolidated Drain Downstream location between 2000 (preconstruction) and 2002 (post-construction). The amount of time that the levels in the drain were above the average level for the July to September period were compared between 2000 and 2002. That percentage rose from forty-five to fifty percent, indicating a decrease in flashiness. A stable channel is expected to have fifty percent of its levels above the average level, and fifty percent below (Booth et al, 2003).

Pollutant removal efficiency of the treatment system was assessed by comparing the annual loading between the Sediment Forebay Inlet and the Wetland Outlet (Table A-4). The EMCs (Tables A-5 and A-6) through the system also provide insight into the way the system performs. Annual loading at the Wetland Outlet was lower than that at the Sediment Forebay Inlet for all parameters except TKN. Cadmium, lead and zinc were measured below the detection levels, indicating that they were effectively removed by the system. Annual loads of TP were not calculated for the Wetland Outlet, as the summer sampling results were not belived to be indicative of year-round loading. However, TP loads estimated for the July 29 to September 9,

2002 period (the period of continuous Wetland Outlet flow) were greater at the Wetland Outlet than the Sediment Forebay Inlet. This indicates that anoxic conditions during the summer months may be causing an export of TP from the system. However, the TP annual loading at the Consolidated Drain Downstream location is 150 pounds lower than that for the Sediment Forebay Inlet and 70 pounds lower than that for the same station during the pre-construction phase, indicating that the elevated TP levels exiting the Wetland Outlet during the summer may not be affecting Portage Creek on an annual basis.

Hydraulic flashiness was compared between the Sediment Forebay Inlet and the Wetland Outlet by evaluating flow during the period of each event at these locations, i.e., the period of time for the same volume of water to flow through each structure. Events at the Sediment Forebay lasted between 6 and 120 hours, with the majority of the flow occurring within the first 3 hours of the event. (As the forebay filled, water backed up into the delivery pipe. The stormwater discharge then slowly drained from the pipe into the forebay, following the initial flush of water.) Wetland Outlet events could last over 260 hours. The detention time for the wetland was five to ten days. Attachment A-5 compares the hydrographs of the Sediment Forebay Inlet and the Wetland Outlet for the July 29 to September 9, 2002 period.

Sediment Forebay Inlet annual loading has increased in comparison to the Upstream preconstruction station likely due to the removal of three small detention basins in the watershed. Stormwater from the 76-acre Crossroads Mall and two other small shopping plazas is now allowed to directly enter the Sediment Forebay, which receives a higher annual load of all parameters as a result (Table A-4). However, due to the TSS concentrations observed exiting the wetland and the calculated annual treatment wetland load of 30,341 pounds (compared to 110,956 pounds at the Sediment Forebay Inlet) it is evident that the system effectively removes 73% of the TSS from the incoming stormwater before returning it to the Consolidated Drain and ultimately to Portage Creek. Similarly, based on annual loading estimates (Table A-4), 83% of the nitrate+nitrite-N and 87% of the copper entering the system is removed.

The anoxic conditions in the wetland may also explain the increase in TKN loading from the system. The Wetland Outlet flow has been visually observed to be high in tannins, indicating that organic materials are exiting the system. The EMCs for TKN (Table A-6) indicate that its concentration rises as water moves through the system. However, the TKN annual load calculated for the Consolidated Downstream location is lower than that for the Sediment Forebay Inlet and for the same station during the preconstruction phase, indicating that this parameter's potential load to Portage Creek is being treated. The EMCs also indicate that TSS, nitrate+nitrite-N and the heavy metals decrease through the system.

A.5 Conclusions

The primary objective of the sampling program was to determine the extent to which the new stormwater treatment facility protects Portage Creek from deleterious stormwater pollutant loads and high velocity flows. On-site observations and the recorded discharge volumes at the Wetland Outlet indicate that both objectives are being met. In contrast to the pre-construction flows and the new Sediment Forebay Inlet, the Wetland Outlet produces a slow, steady release of treated stormwater. (See hydrographs in Attachment A-5.)

Post-construction TP and TKN concentrations discharging from the wetland were observed to be greater than inflowing levels during the summer months. It is presumed that the binding sites on the existing wetland soils are saturated, given that these existing wetlands have been active for approximately 50 years since the area was previously utilized for celery (muck) farming. Anoxic conditions in the wetland may allow any sediment bound phosphorus to be released, and primary productivity and decomposition allows organic materials to be released. In contrast, measured TP concentrations exiting the wetland in the early spring (0.12 mg/L) were much lower than the Wetland Outlet TP summer EMC (0.34 mg/L) and than the Sediment Forebay Inlet TP EMC (0.16 mg/L). Further monitoring is necessary to determine whether the phosphorus sequestered in the wetland during the winter and spring is all released during the summer growing season and whether the discharge of organic materials during the summer months occurs year-round. These limited data do not clearly indicate whether there was a net annual release of phosphorus and TKN from the system; and, although concentrations leaving the wetland were high, concentrations measured at the Downstream location indicate that these loads to Portage Creek have decreased following the installation of the new treatment system.

It is recommended that year-round (seasonal) monitoring at the Sediment Forebay Inlet, Wetland Outlet and Downstream Consolidated Drain stations be conducted to continually assess the annual treatment efficiency of the system. It is also suggested that monitoring of the Schuring Drain upstream and downstream of its confluence with a private drain in the lower portions of this drainage be conducted to assess whether loading is occurring from watershed runoff or bed scour of the reshaped channel. The latter situation is expected to remedy itself as establishment of mature vegetational communities stabilize the channel banks and bottom.

Overall, monitoring suggests that the primary objectives of water quality treatment and mitigation of hydraulic flashiness of the Consolidated Drain have been achieved by the treatment system. Growth of replanted and pioneer vegetation has proceeded with each growing season, and the overall area now sustains healthy stands of vegetation within, adjacent to and surrounding the stormwater treatment system. It is also evident from monitoring data that the newly connected Schuring Drain has notable impacts on the Consolidated Drain due to unmitigated stormwater runoff from this southern drainage system. Due to the low-lying relief in this area and the volume limitations for the Stormwater Quality and Trailways Facility, the Schuring Drain could not be connected to this system. Upstream areas of the Schuring Drain Watershed may require other forms of mitigation for smaller catchment areas to address these unmitigated loads.

A.6 References

Booth, D.B. et al. Center for Water and Watershed Studies, University of Washington. *Strategies and Limitations for Urban Watershed Planning*. Presentation at Urban Storm Water: Enhancing Programs at the Local Level. Chicago, IL. February 17-20, 2003.

Debo, T.N. and A.J. Reese. 1995. Municipal Stormwater Management. Lewis Publishers, Ann Arbor, MI.

Kieser & Associates/URS Greiner Woodward Clyde (K&A/URS). 1999. *Consolidated Drain No.1 Feasibility Study: An Integrated Approach Focusing on Environmental & Public Benefits for the City of Portage.* Prepared for the City of Portage, Department of Transportation & Utilities.

Kieser & Associates. 2001. *Non-Point Source Modeling of Phosphorus Loads in the Kalamazoo River/Lake Allegan Watershed for a TMDL.*

Wayne County Rouge River National Wet Weather Demonstration Project (Wayne County). 1998. *Urban Storm Water Quantification Protocols.* Prepared for the State of Michigan, Water Quality Trading Workgroup.

Figure A-1. Extent of Consolidated Drain and Schuring Drain Watersheds in the City of Portage.

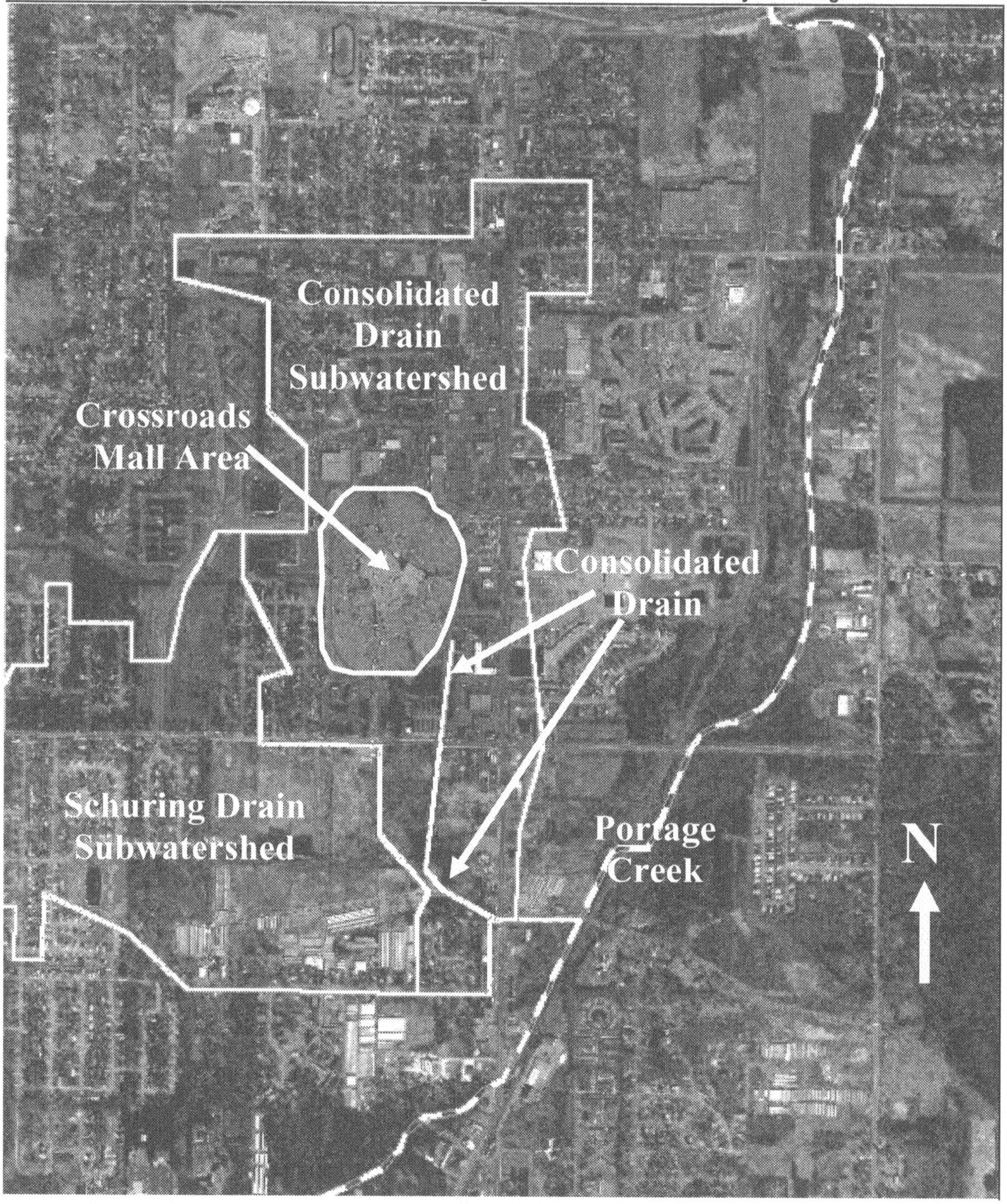

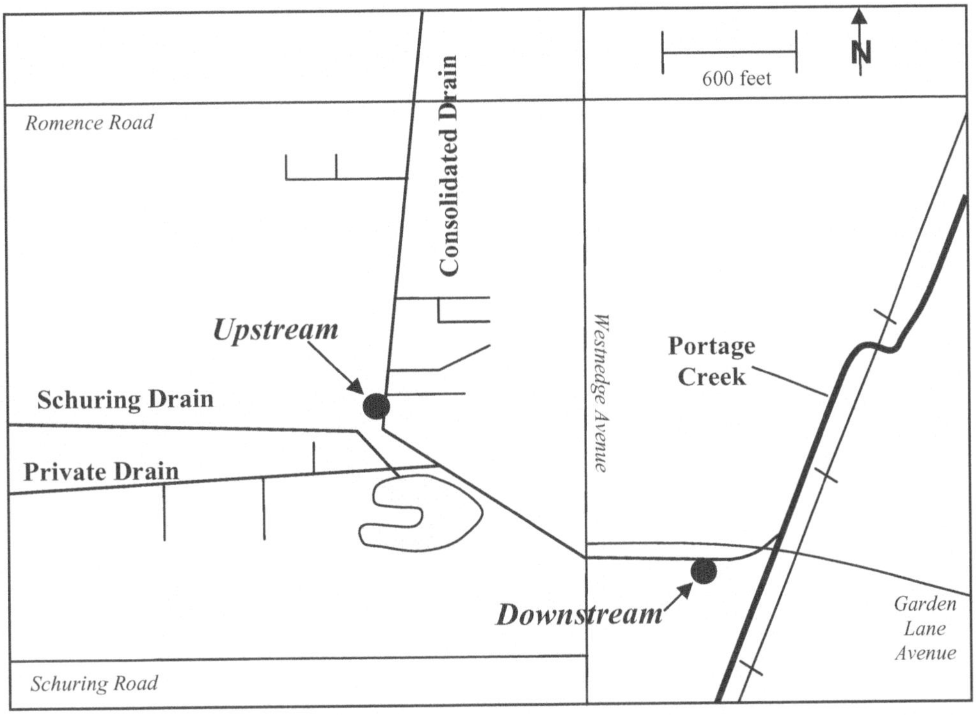

Figure A-2. Preconstruction Consolidated Drain configuration and K&A automated sampling stations, 1999.

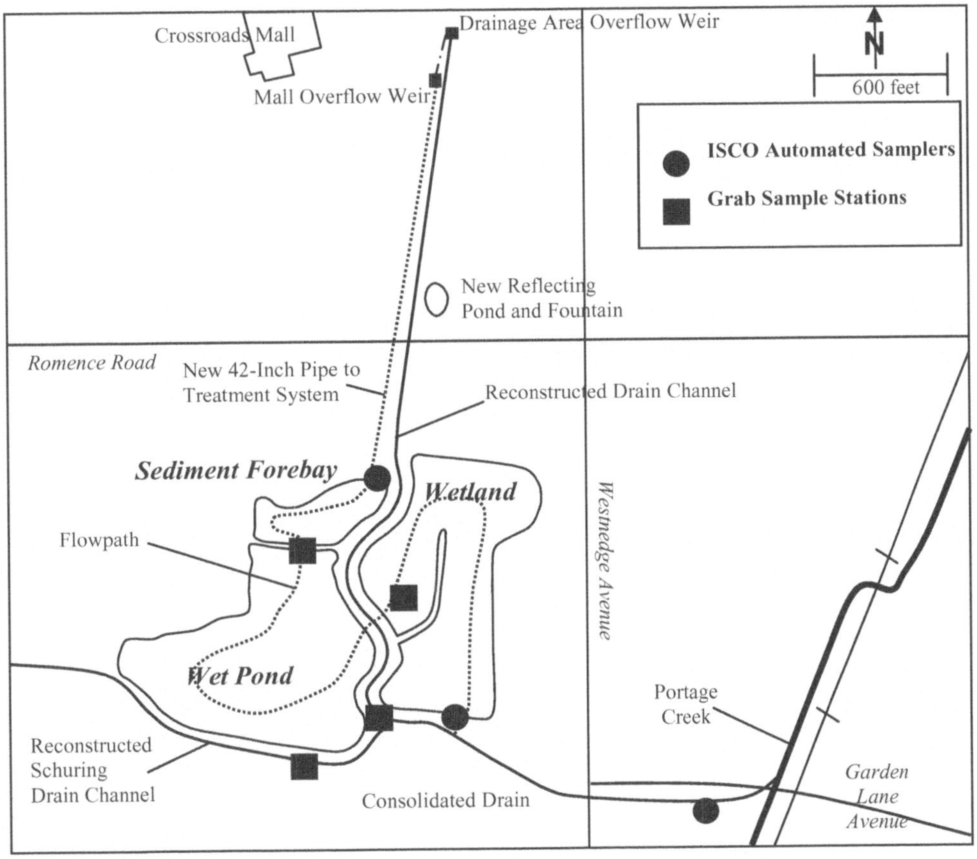

Figure A-3. Post-construction Consolidated Drain configuration and K&A automated sampling stations, 2002.

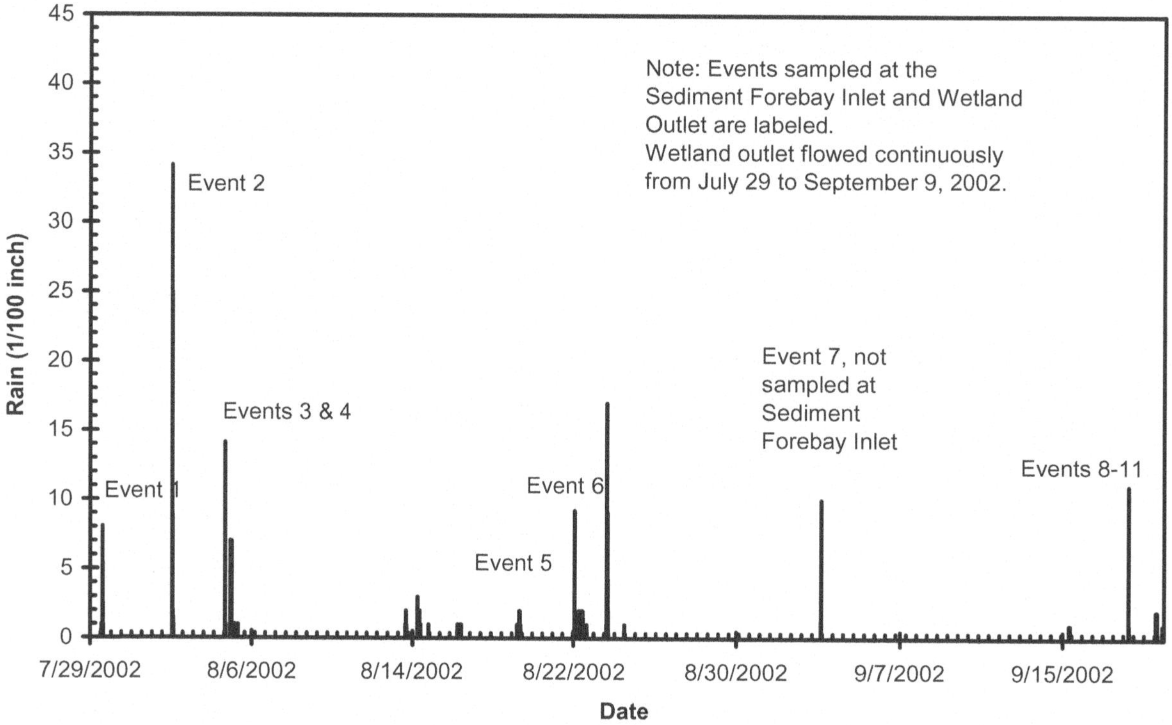

Figure A-4. Rainfall hyetograph from July 29 to September 20, 2002.

Attachment A-1
Stage-Discharge Curves for the Consolidated Drain

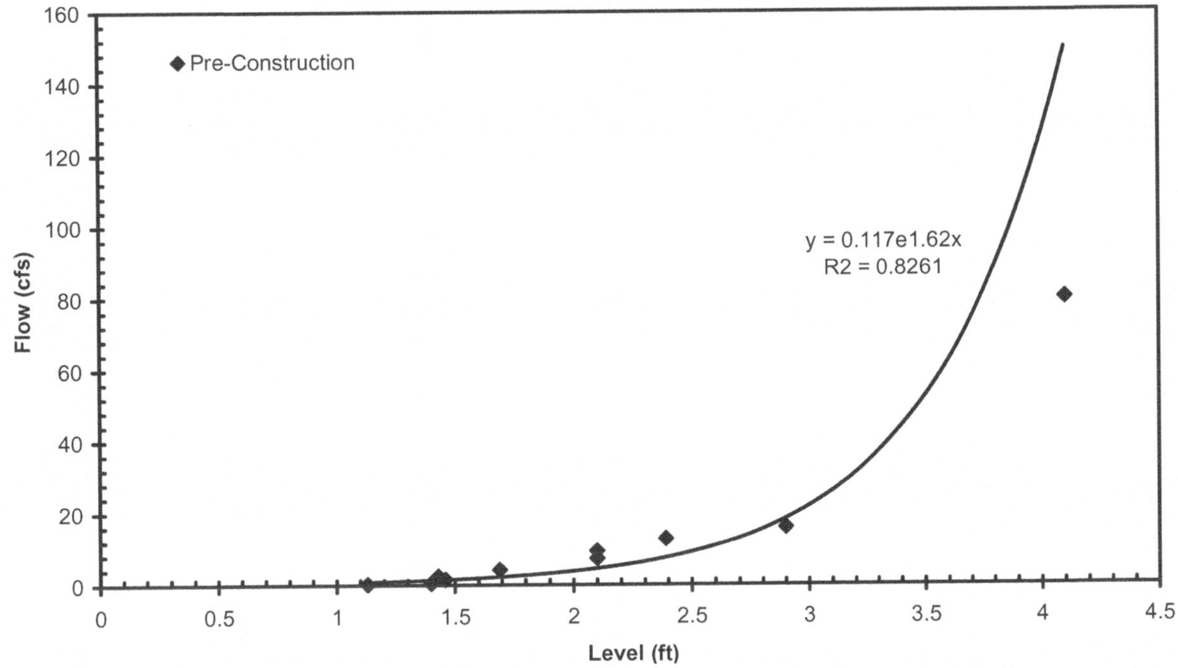

Figure A-5. Preconstruction Portage Consolidated Drain upstream exponential stage discharge relationship.

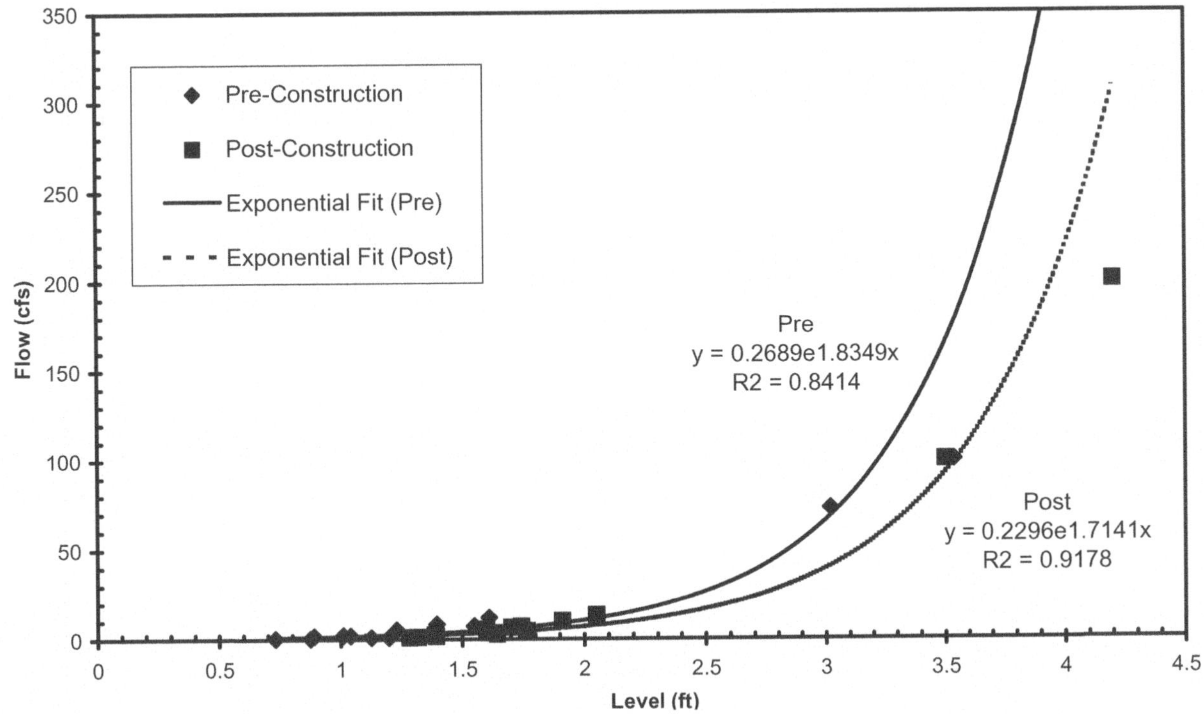

Figure A-6. Pre- and Post-Construction Portage Consolidated Drain Downstream exponential stage discharge relationships.

Attachment A-2
Comparisons of Total Phosphorus Concentrations from Three Analytical Laboratories

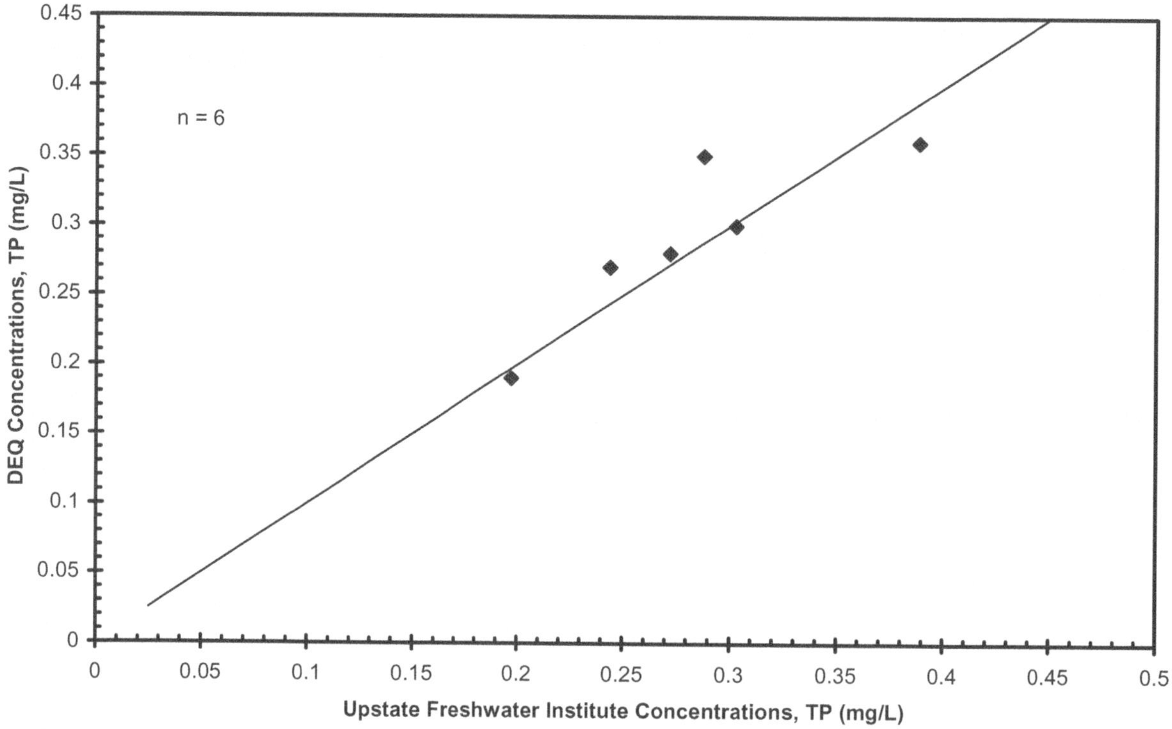

Figure A-7. MDEQ TP concentrations vs. Upstate Freshwater Institute TP concentrations.

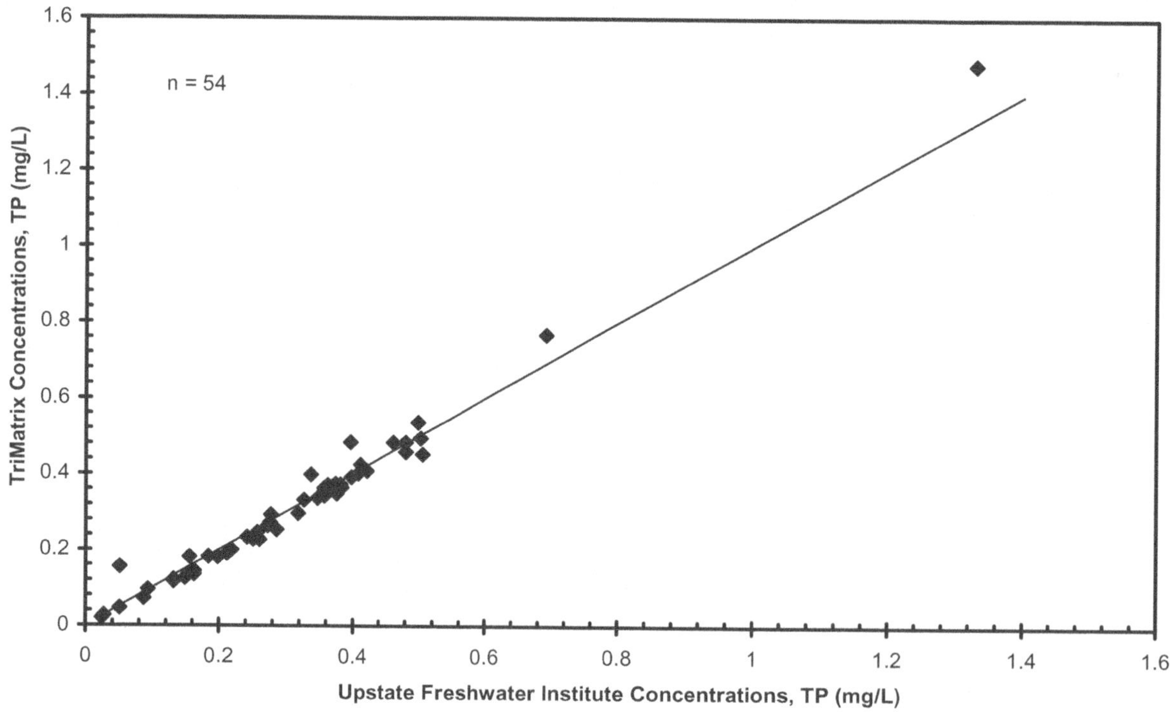

Figure A-8. Tri-Matrix TP concentrations vs. Upstate Freshwater Institute TP concentrations.

Attachment A-3
Preconstruction Load vs. Rainfall Relationships and Load Totals for Events at the Upstream and Downstream Consolidated Drain Stations

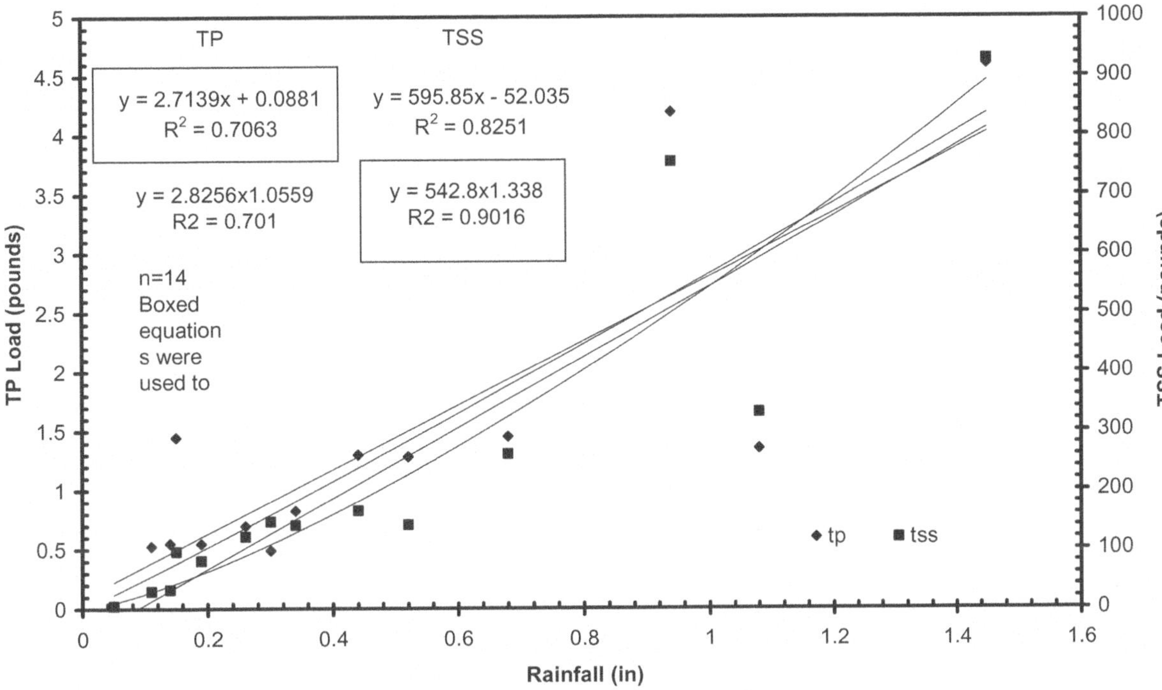

Figure A-9. TP and TSS loads vs rainfall-Condrain Upstream Preconstruction.

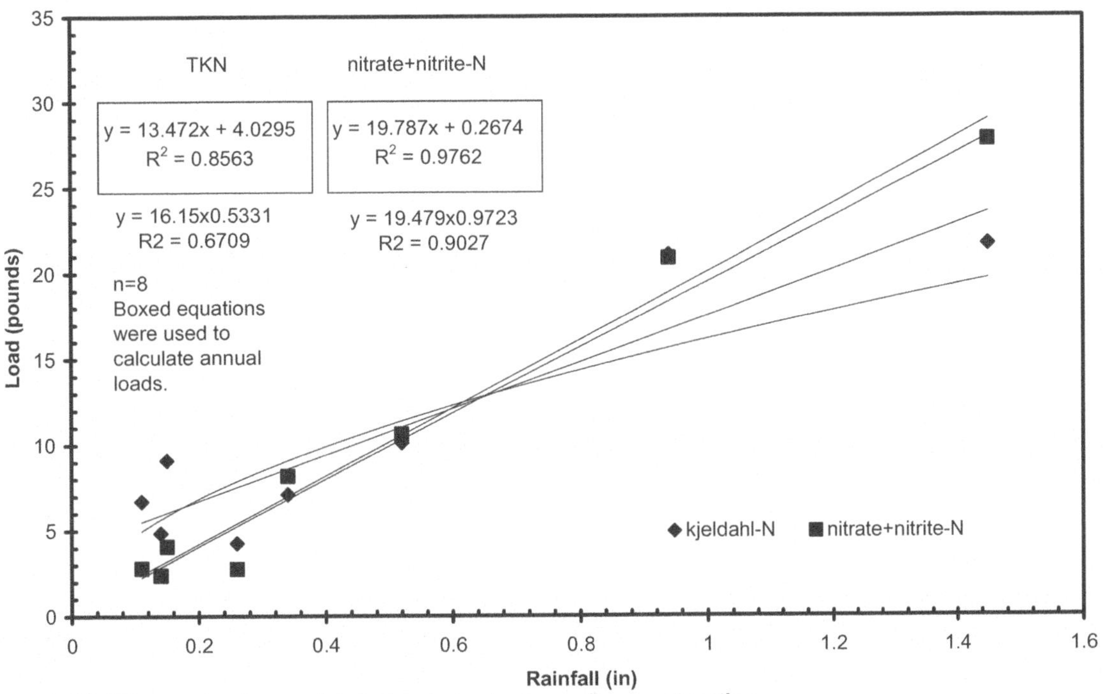

Figure A-10. Nitrogen loads vs rainfall-Condrain Upstream Preconstruction.

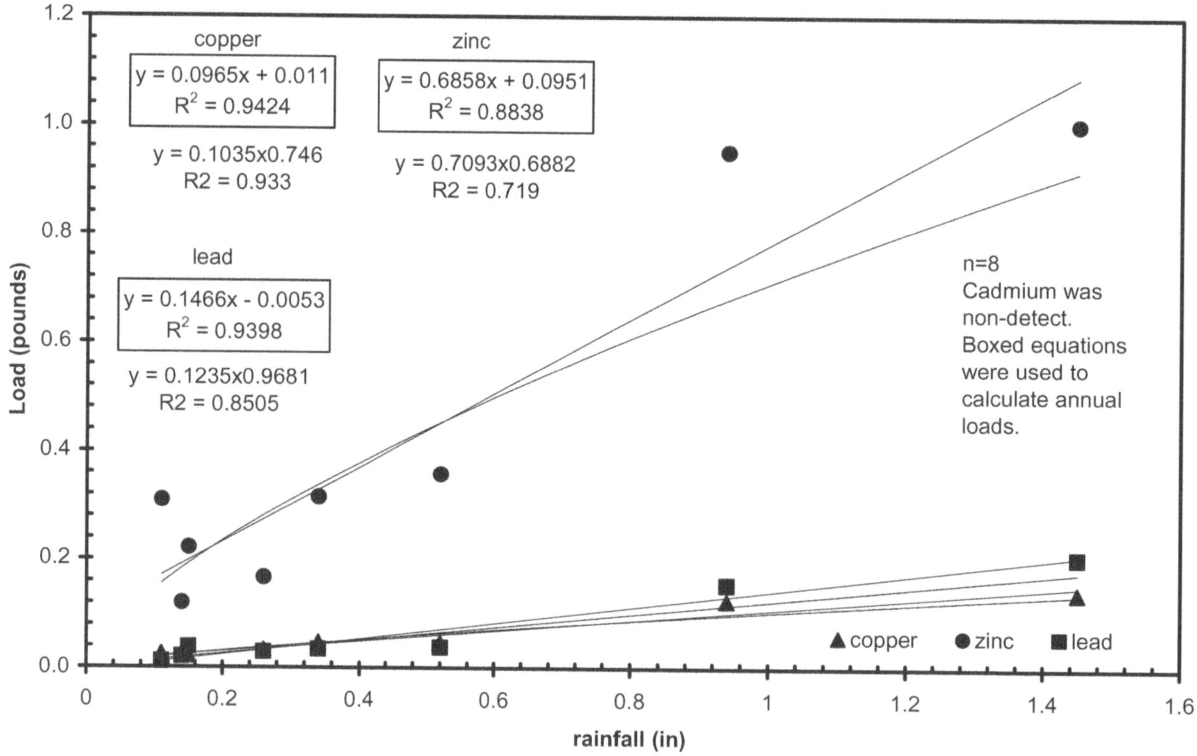

Figure A-11. Metal loads vs rainfall-Condrain Upstream Preconstruction.

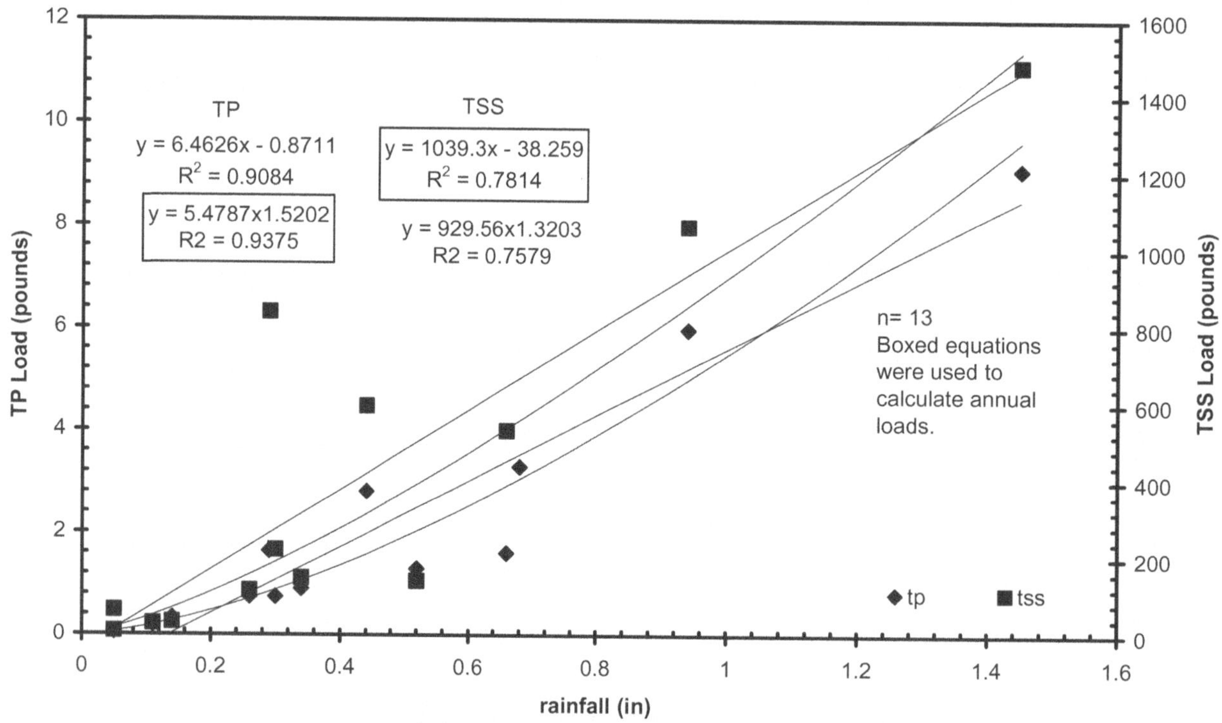

Figure A-12. TP and TSS loads vs rainfall-Condrain Downstream Preconstruction.

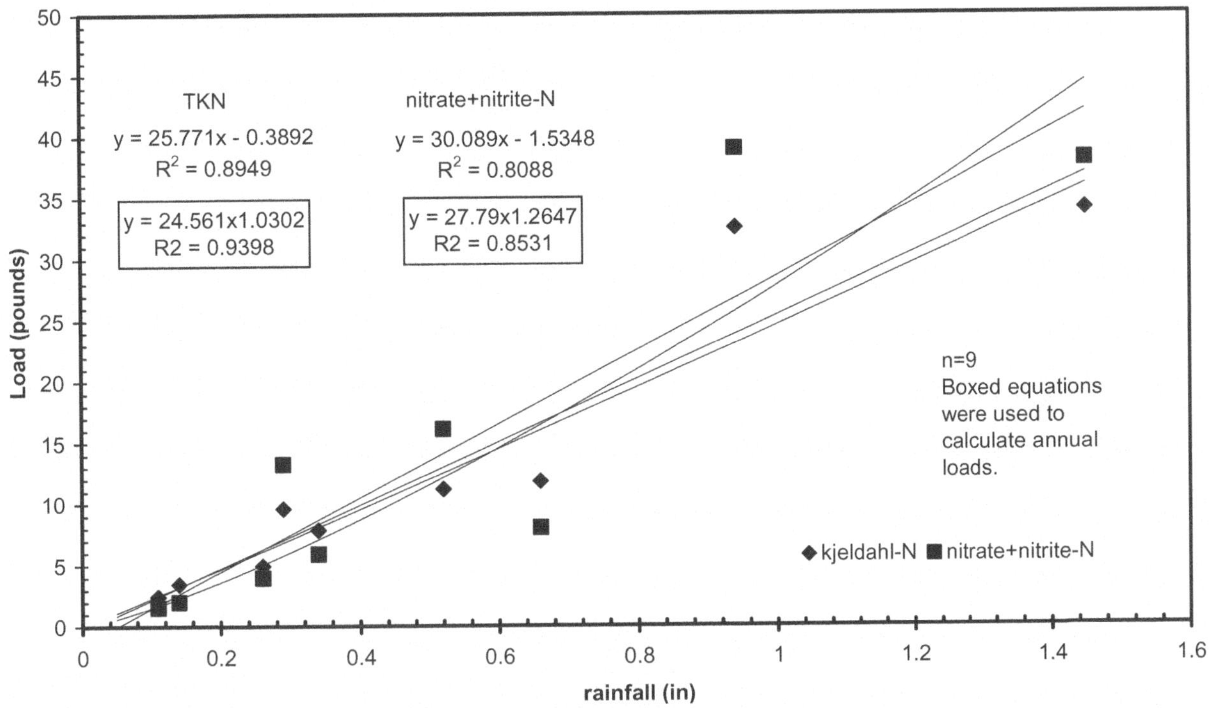

Figure A-13. Nitrogen loads vs rainfall-Condrain Downstream Preconstruction.

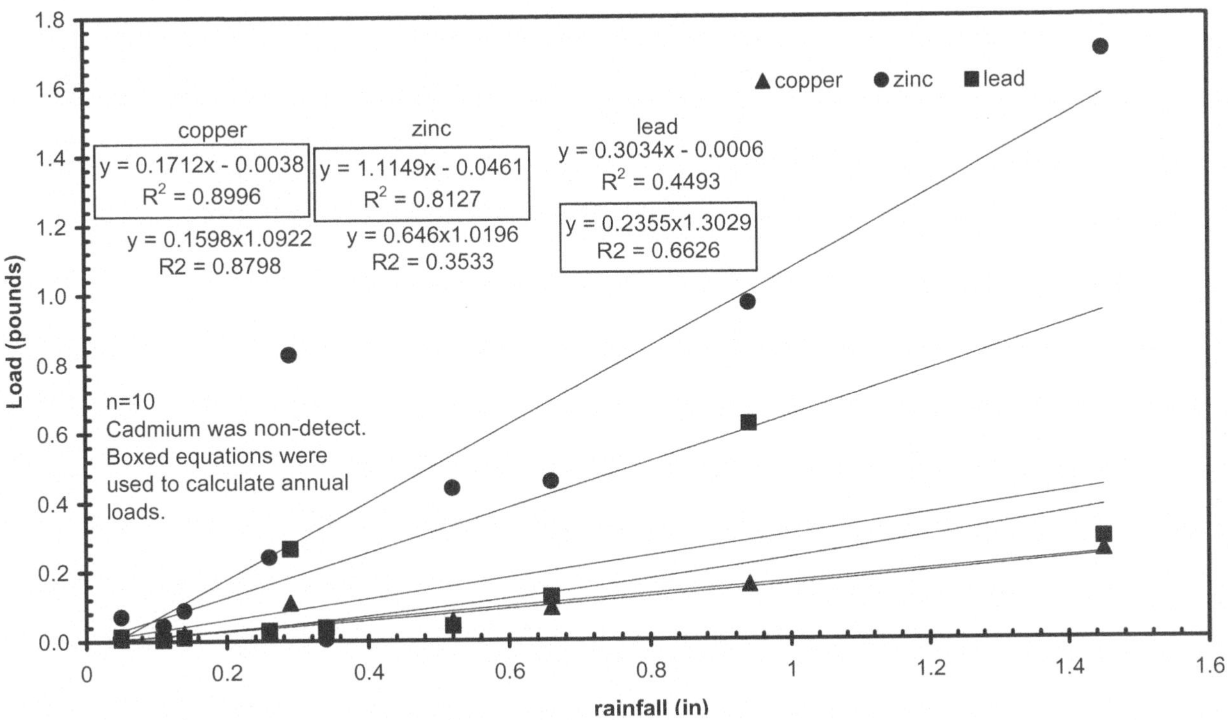

Figure A-14. Metal loads vs rainfall-Condrain Downstream Preconstruction.

Table A-7. Upstream automated sampler preconstruction loading (in pounds).

Event Date	volume (cubic feet)	rainfall (inches)	nitrite-N	SRP	TSS	no3+ no2-N	ammonia-N	kjeldahl-N
11/02/99	162,962	1.08		0.354	331.06	8.610	3.788	
03/15/00	10,826	0.05			4.71			
03/16/00	71,570	0.3		0.021	146.67	6.865	2.271	
05/09/00	10,1167	0.44			165.27			
05/10/00	202,110	0.68			261.05			
07/10/00	100,696	0.14	0.239	0.218	32.00	2.394	0.656	4.857
07/21/00	45,021	0.11	0.323	0.107	29.34	2.816	2.425	6.705
08/02/00	169,619	0.15	0.476	0.839	96.61	4.087	2.779	9.103
08/05/00	181,568	0.52	0.637	0.403	141.37	10.611	2.247	10.120
08/17/00	119,085	0.34	0.300	0.331	140.56	8.181	2.064	7.099
09/11/00	312,947	0.94	0.585	1.437	754.74	20.870	4.435	21.063
09/20/00	70,706	0.26	0.195	0.139	121.30	2.764	0.742	4.270
10/03/00	516,696	1.45	0.631	2.382	928.61	27.784	5.863	21.694
10/05/00	58,111	0.19			80.61			

Event Date	TP	arsenic	barium	chromium	copper	lead	zinc
11/02/99	1.350						
03/15/00	0.028						
03/16/00	0.489						
05/09/00	1.298						
05/10/00	1.452						
07/10/00	0.550	0.009	0.119	0.015	0.025	0.020	0.120
07/21/00	0.527	0.007	0.110	0.008	0.025	0.011	0.309
08/02/00	1.445	0.027	0.223	0.019	0.022	0.038	0.221
08/05/00	1.281	0.019	0.198	0.026	0.047	0.038	0.357
08/17/00	0.825	0.019	0.225	0.031	0.047	0.034	0.315
09/11/00	4.192	0.043	0.530	0.083	0.125	0.155	0.952
09/20/00	0.695	0.010	0.134	0.021	0.034	0.029	0.166
10/03/00	4.604	0.039	0.441	0.060	0.141	0.205	1.002
10/05/00	0.545						

Table A-8. Downstream automated sampler preconstruction loading (in pounds).

Event Date	volume (cubic feet)	rainfall (inches)	nitrite-N	SRP	TSS	no3+ no2-N	ammonia-N	kjeldahl-N
03/15/00	23,521	0.05			8.41			
03/16/00	102,207	0.3		0.020	221.04	6.956	2.838	
05/09/00	120,504	0.44			595.53			
05/10/00	228,547	0.68			5,421.24*			
07/10/00	86,507	0.14	0.181	0.114	33.12	2.015	0.489	3.464
07/21/00	41,575	0.11	0.173	0.056	29.33	1.557	0.725	2.447
08/05/00	229,464	0.52	0.860	0.354	140.85	16.113	2.223	11.190
08/17/00	160,013	0.34	0.116	0.221	147.96	5.915	1.694	7.869
09/11/00	470,216	0.94	1.049	2.212	1,063.11	38.990	6.314	32.517
09/20/00	104,877	0.26	0.280	0.131	115.70	3.977	0.987	4.941
10/03/00	657,559	1.45	0.920	3.560	1,481.94	38.066	8.086	34.025
02/08/01	23,839	0.05			63.25			
04/06/01	118,373	0.29	0.269	0.197	839.32	13.254	2.116	9.609
06/12/01	200,748	0.66	0.306	0.345	531.14	8.019	2.872	11.815

Event Date	TP	arsenic	barium	chromium	copper	lead	zinc
03/15/00	0.043						
03/16/00	0.743						
05/09/00	2.801						
05/10/00	3.284						
07/10/00	0.332	0.009	0.177		0.022	0.011	0.089
07/21/00	0.189	0.005	0.142	0.003	0.008	0.004	0.045
08/05/00	1.289	0.023	0.405	0.031	0.057	0.042	0.440
08/17/00	0.899	0.027	0.438	0.024	0.039	0.037	0.005
09/11/00	5.963	0.059	0.757	0.105	0.157	0.622	0.971
09/20/00	0.746	0.014	0.296	0.021	0.029	0.030	0.241
10/03/00	9.093	0.133	0.774	0.086	0.255	0.290	1.699
02/08/01		0.003	0.124	0.004	0.007	0.013	0.071
04/06/01	1.634	0.030	0.409	0.078	0.109	0.265	0.827
06/12/01	1.599	0.020	0.308	0.040	0.093	0.124	0.458

Attachment A-4
Post-construction Load vs. Rainfall Relationships and Load Totals for Events at the Consolidated Drain Downstream Station

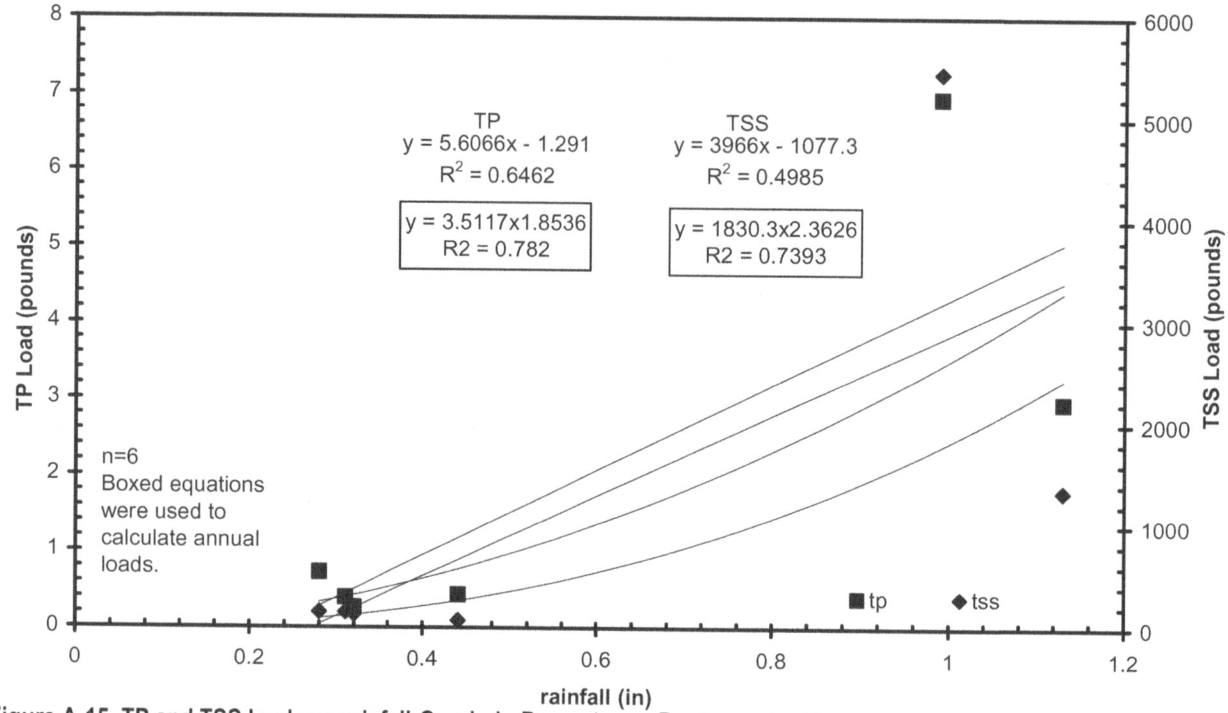

Figure A-15. TP and TSS loads vs rainfall-Condrain Downstream Post-construction.

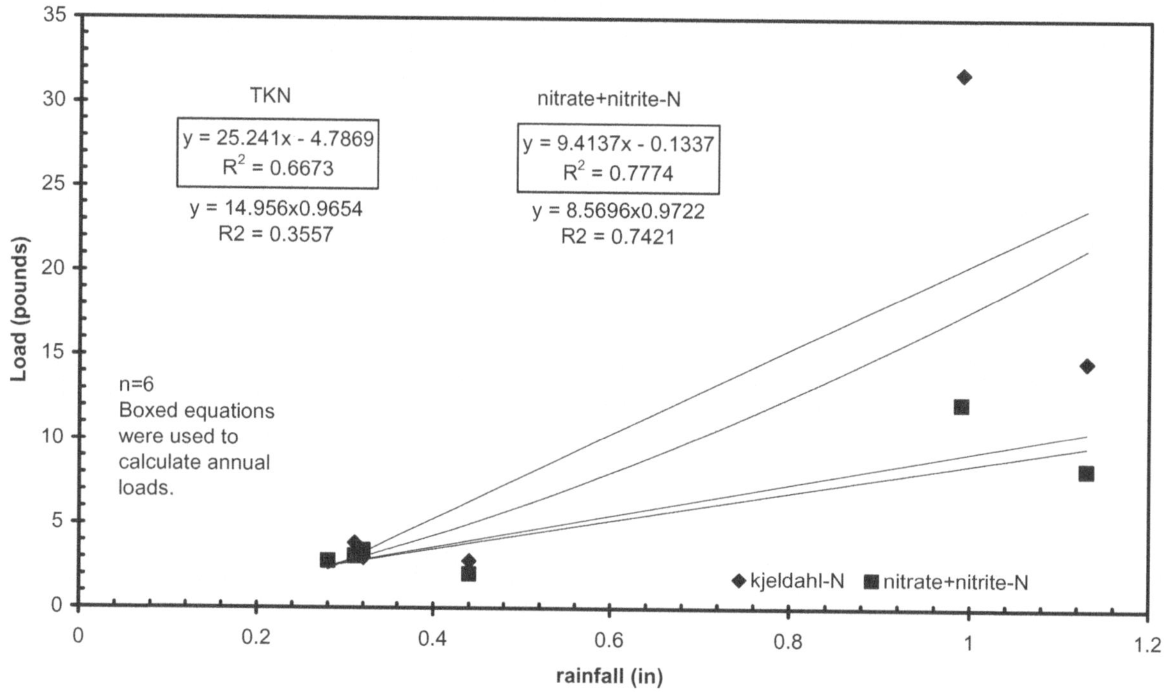

Figure A-16. Nitrogen loads vs rainfall-Condrain Downstream Post-construction.

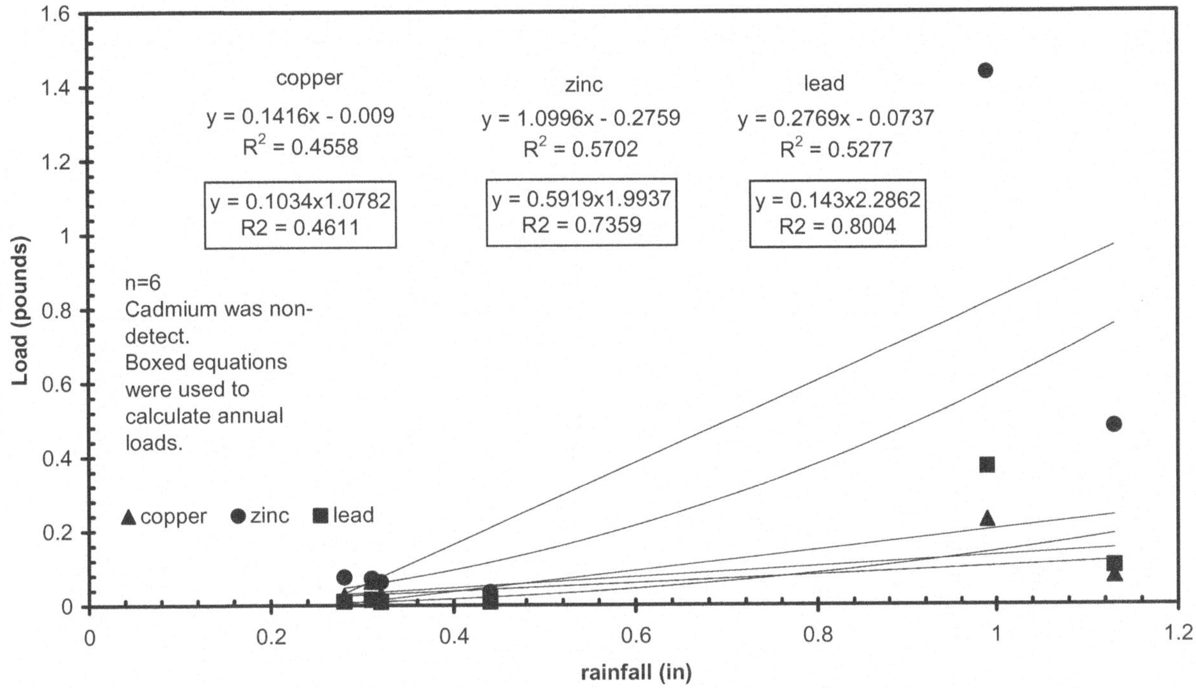

Figure A-17. Metal loads vs rainfall-Condrain Downstream Post-construction.

Table A-9. Downstream automated sampler post-construction loading (in pounds).

Event Date	volume (cubic feet)	rainfall (inches)	TSS	ammonia-N	no2+no3-N	nitrite-N	kjeldahl-N	SRP
03/08/02	296,748	0.23 + snow melt	3,980.85	14.135	23.446	0.2831	15.599	1.372
05/08/02	219,603	1.13	1,344.43	2.2616	8.3152	0.2612	14.7073	0.4670
07/22/02	64,853	0.32	130.04	0.387	3.434	0.154	2.981	0.054
07/29/02	56,251	0.31	150.88	0.200	3.081	0.085	3.852	0.085
08/02/02	280,211	0.99	5,453.64	2.613	12.207		31.785	
08/04/02	61,720	0.28	150.79	0.330	2.790	0.102	2.732	0.270
09/20/02	62,775	0.44	68.31	0.440	2.036	0.070	2.810	0.107

Event Date	**TP**	**arsenic**	**barium**	**chromium**	**copper**	**lead**	**zinc**
03/08/02	2.972	0.059	1.168	0.132	0.276	0.373	1.386
05/08/02	2.954	0.0375	0.4932	0.0450	0.077	0.104	0.480
07/22/02	0.261	0.007	0.288	0.006	0.029	0.009	0.062
07/29/02	0.395	0.008	0.184	0.007	0.064	0.016	0.072
08/02/02	6.946	0.078	1.269	0.117	0.620	0.372	1.435
08/04/02	0.723	0.008	0.189	0.005	0.026	0.010	0.076
09/20/02	0.431	0.007	0.255	0.005	0.015	0.009	0.035

Attachment A-5

Post-construction Hydrographs, Load vs. Rainfall Relationships and Load Totals for Events at the Sediment Forebay Inlet and Wetland Outlet Stations

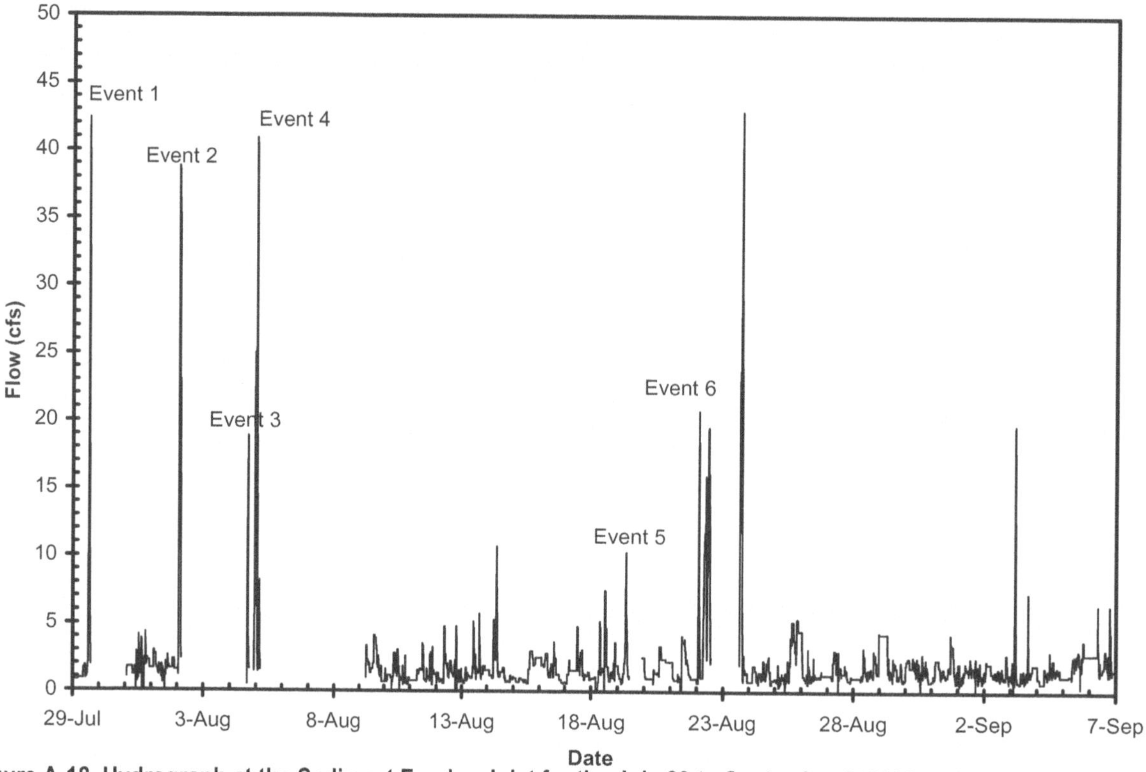

Figure A-18. Hydrograph at the Sediment Forebay Inlet for the July 29 to September 9, 2002 period.

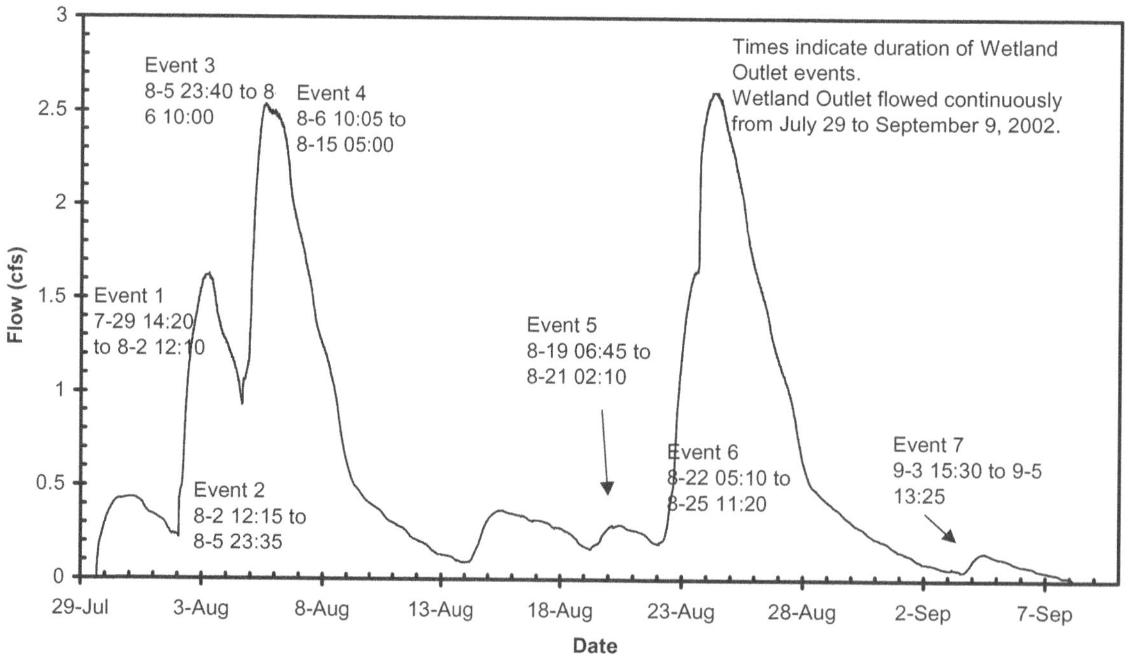

Figure A-19. Hydrograph at the Wetland Outlet for the July 29 to September 9, 2002 period.

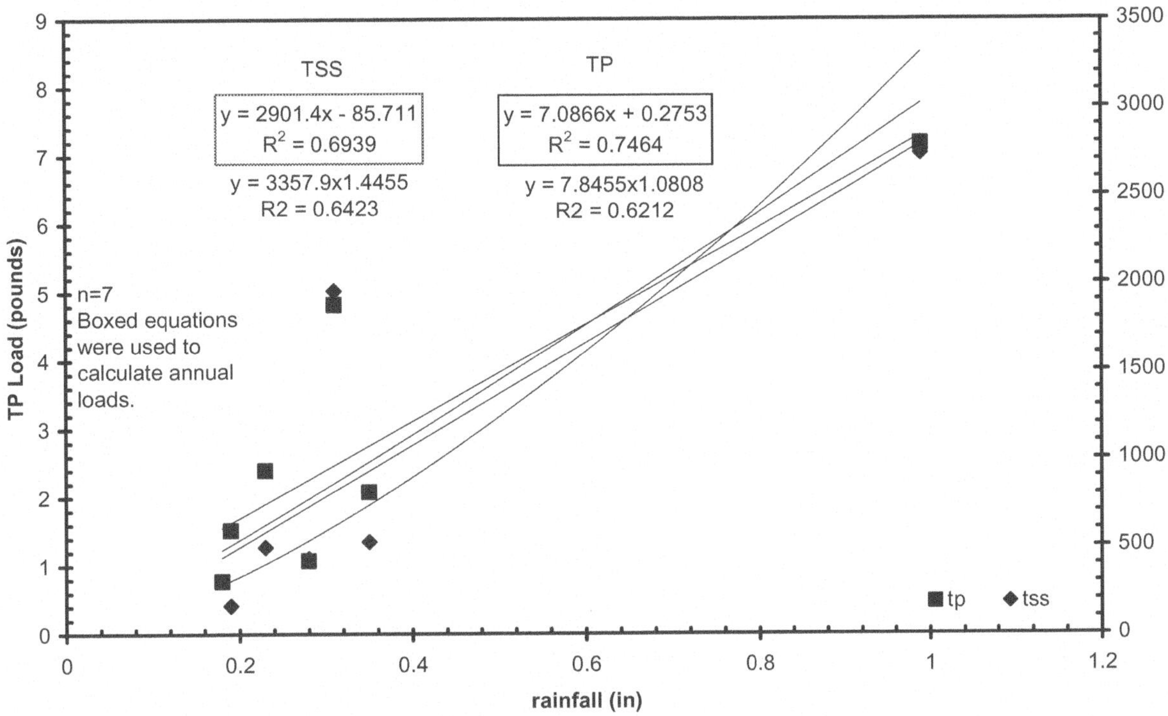

Figure A-20. TP and TSS loads vs rainfall-Sediment Forebay Inlet.

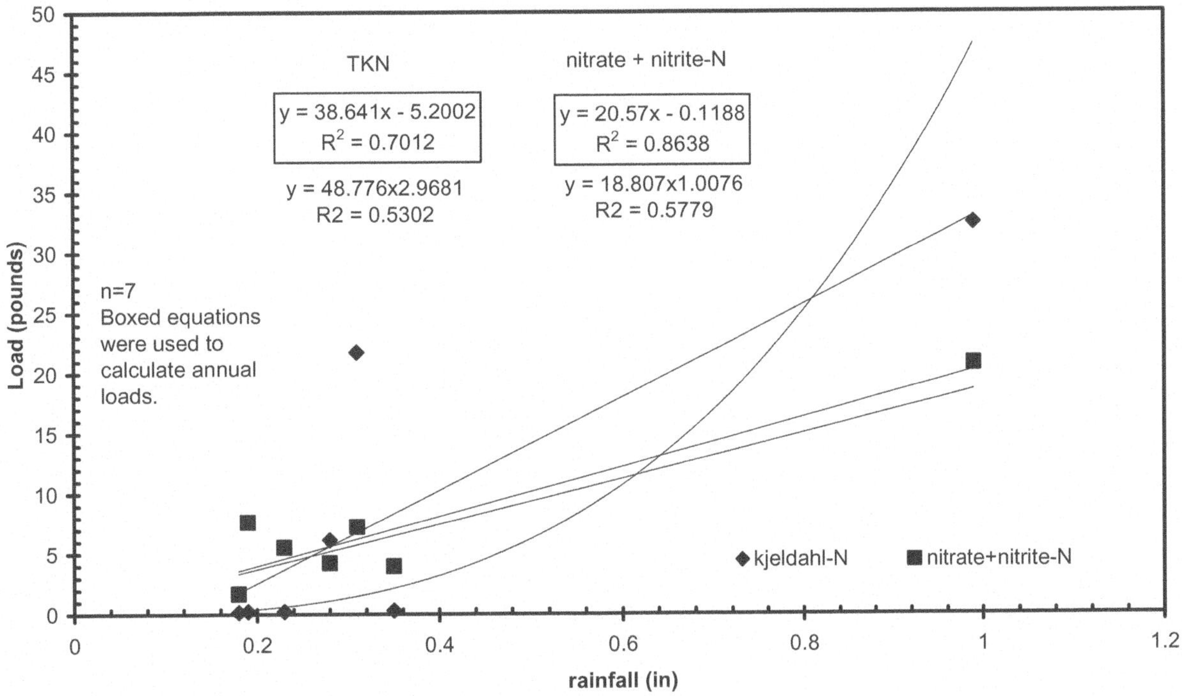

Figure A-21. Nitrogen loads vs. rainfall-Sediment Forebay Inlet.

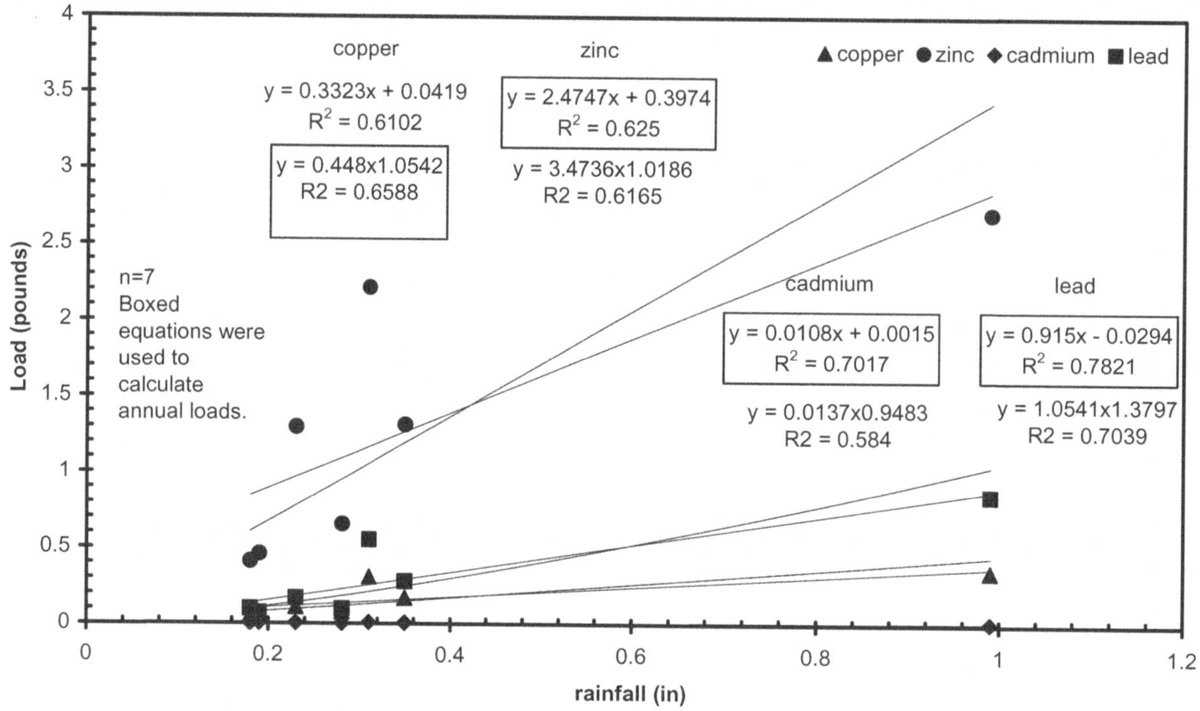

Figure A-22. Metal loads vs rainfall-Sediment Forebay Inlet.

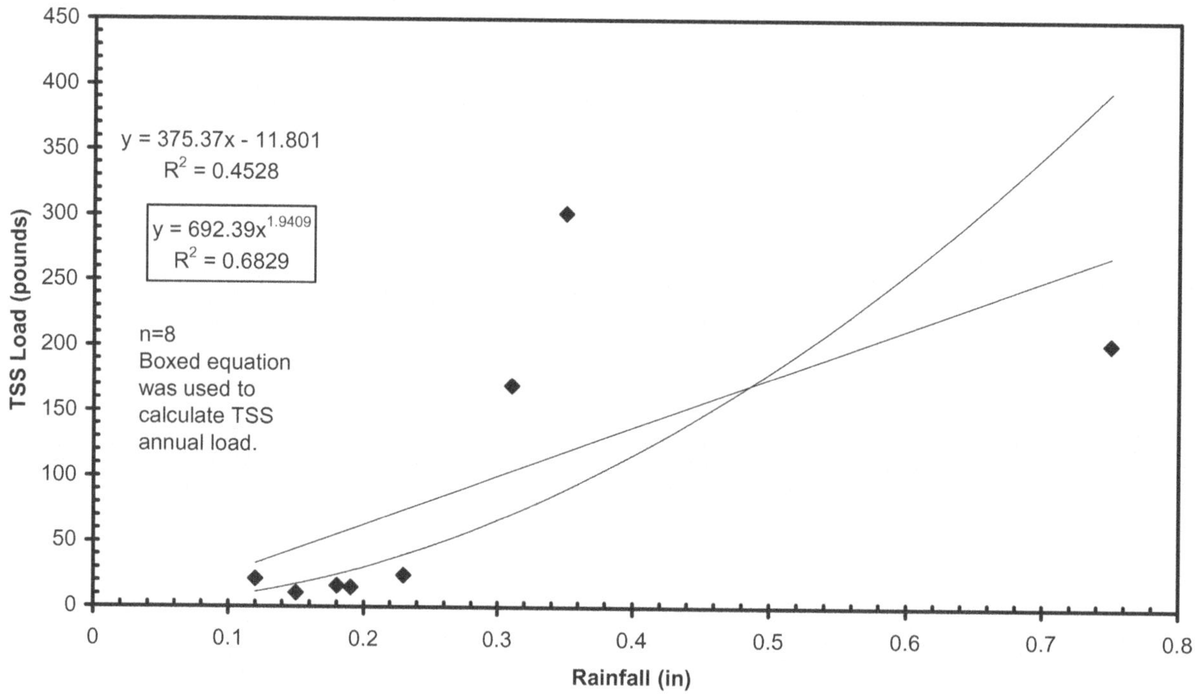

Figure A-23. TSS loads vs rainfall-Wetland Outlet.

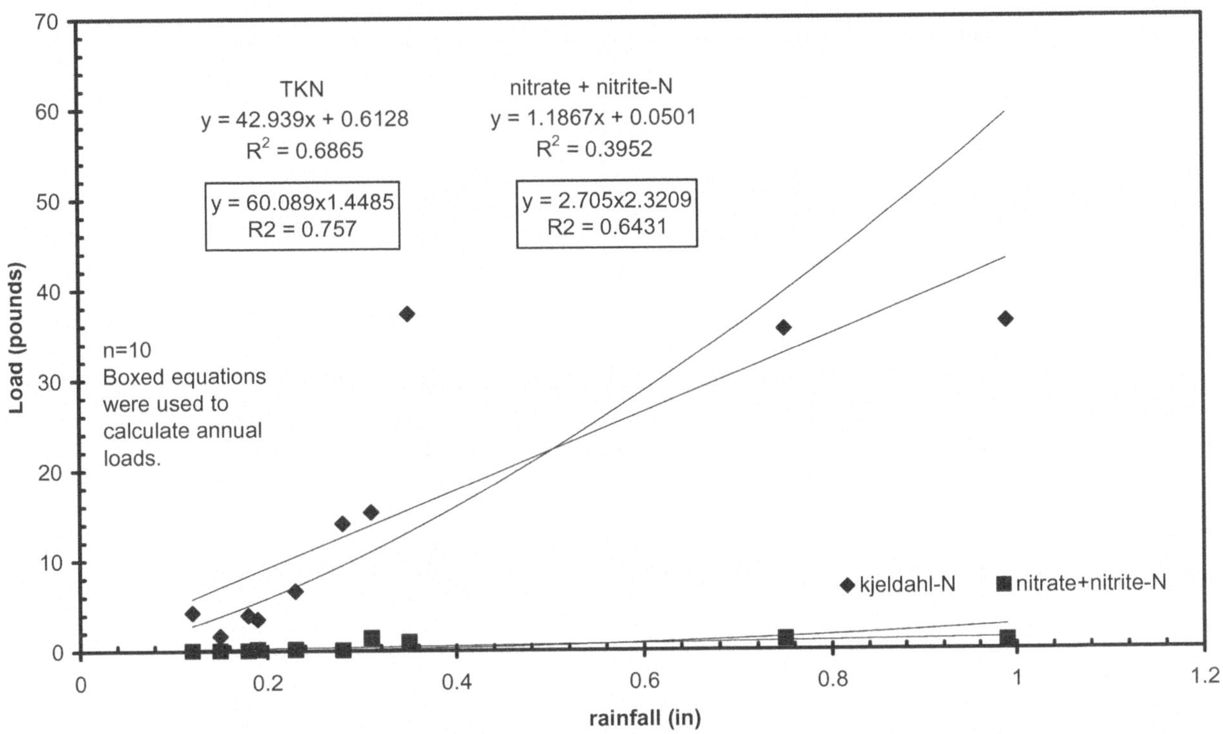

Figure A-24. Nitrogen loads vs. rainfall-Wetland Outlet.

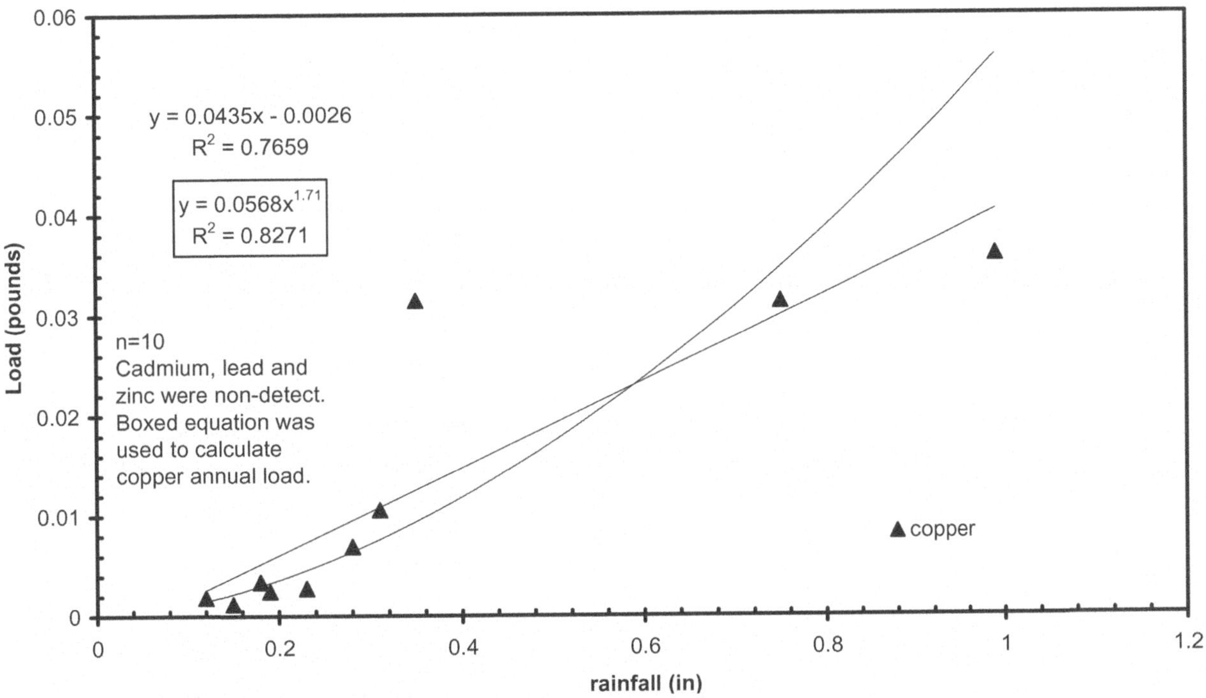

Figure A-25. Metal loads vs rainfall-Wetland Outlet.

Table A-10. Sediment Forebay Inlet salt and solid loading (pounds).

rainfall	event #	hardness	chloride	sodium	TSS
0.31	1	994.66	232.79	176.03	1,955.16
0.99	2	1,081.12	124.82	120.23	2,733.90
0.28	3	231.38	51.85	36.15	430.55
0.75	4	1,556.81	175.57	146.94	1,470.55
0.19	5	596.80	216.49	128.14	163.83
0.35	6	697.06	84.55	39.39	525.88
0.23	8	654.04	255.44	136.59	494.71
0.18	10	200.79	46.64	30.84	284.45
0.26	11	1,011.26	131.38	100.48	2,047.66

Table A-11. Wetland Outlet salt and solid loading (pounds).

rainfall	event #	hardness	chlorine	sodium	TSS
0.31	1	1,698.15	933.06	646.79	169.67
0.99	2	3,466.61	1566.46	974.91	
0.28	3	575.36	575.36	136.12	
0.75	4	3,050.76	863.68	574.52	202.65
0.19	5	402.76	114.46	76.06	15.03
0.35	6	3,566.65	941.78	624.64	301.83
0.15	7	187.19	32.53	21.59	10.53
0.23	8	565.55	155.28	103.02	24.34
0.12	9	371.33	122.87	74.12	21.20
0.18	10	375.52	120.41	72.27	16.19

Table A-12. Sediment Forebay Inlet nutrient loading (pounds).

event #	ammonia-N	no3+no2-N	nitrite-N	kjeldahl-N	SRP	TP
1	1.255	7.263	0.493	21.779	0.549	4.826
2	13.302	20.800		32.501		7.171
3	1.090	4.255	0.172	6.153	0.120	1.077
4	8.499	10.030	1.841	27.029	0.566	5.461
5	0.117	7.665	0.234	8.776	0.234	1.521
6	1.464	3.981	0.309	8.290	0.619	2.083
8	3.086	5.549	0.225	12.805	0.322	2.400
10	0.297	1.697	0.160	4.580	0.077	0.777
11	3.100	10.087	0.893	25.531	0.691	5.754

Table A-13. Wetland Outlet nutrient loading (pounds).

event #	ammonia-N	no3+no2-N	nitrite-N	kjeldahl-N	SRP	TP
1	1.041	1.389		15.339		3.548
2	1.800	0.878		36.272		11.788
3	0.057	0.0570		14.096		2.016
4	1.578	1.123		35.493		7.390
5	0.509	0.176		3.535		0.778
6	4.232	0.916		37.318		10.411
7	0.280	0.012		1.626		0.313
8	1.019	0.138	0.071	6.648		1.502
9	0.500	0.012	0.026	4.260		1.211
10	0.531	0.015	0.021	3.974		1.156

Table A-14. Sediment Forebay Inlet metal loading (pounds).

event #	arsenic	barium	cadmium	chromium	copper	lead	zinc
1	0.076	0.684	0.009	0.341	0.307	0.551	2.213
2	0.101	0.817	0.012	0.248	0.351	0.850	2.708
3	0.013	0.150	0.002	0.066	0.077	0.097	0.653
4	0.079	0.753	0.011	0.160	0.269	0.316	2.109
5	0.037	0.187	0.004	0.028	0.070	0.070	0.456
6	0.032	0.408	0.006	0.296	0.170	0.277	1.309
8	0.044	0.325	0.004	0.071	0.108	0.167	1.295
10	0.024	0.104	0.002	0.047	0.051	0.098	0.408
11	0.078	0.826	0.006	0.410	0.343	0.579	2.560

Table A-15. Wetland Outlet metal loading (pounds).

event #	arsenic	barium	copper
1	0.034	0.329	0.011
2	0.098	0.974	0.036
3	0.017	0.165	0.007
4	0.072	0.840	0.031
5	0.007	0.096	0.002
6	0.074	0.844	0.031
7	0.003	0.044	0.001
8	0.023	0.126	0.003
9	0.008	0.081	0.002
10	0.008	0.080	0.003

APPENDIX B

INSTRUMENTATION AND DATABASE CODING

This appendix contains a map of monitoring locations in the study area. Detailed descriptions of instruments deployed at each monitoring location are included. In addition, database coding is included in the descriptions for future reference and use in analysis. This information provides detail that was condensed in previous chapters and provides documentation of the organization of the long-term dataset.

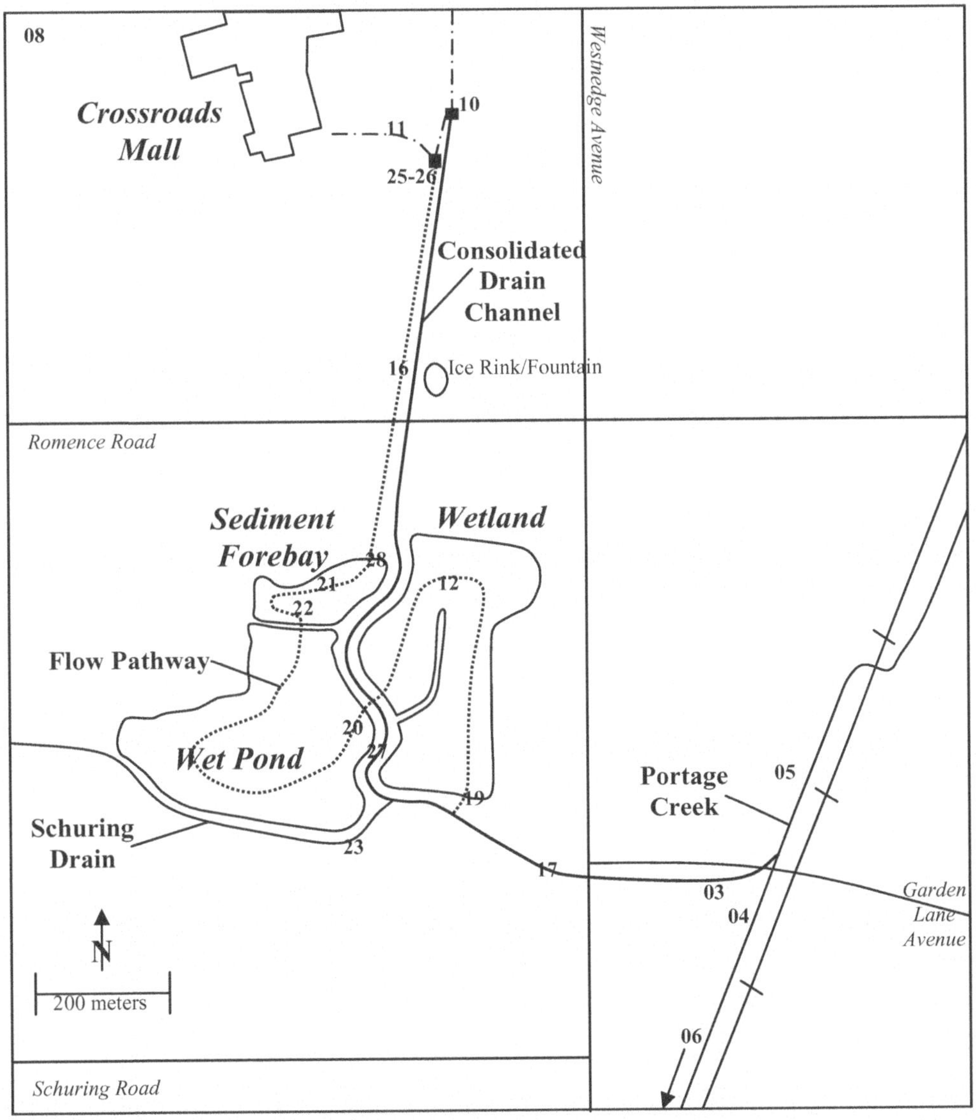

Figure B-1. Post-construction monitoring instrument locations (location numbers correspond with instrument description).

General Database Codes

- r - rain (tips per/5 minutes), 1 tip = 0.01 inch
- a - air parameters, including dew point (ºC), temperature (ºC), absolute humidity (g/m^3), and relative humidity (%)
- slr - solar radiation ($watts/m^2$)
- kbs hourly total MJ - substitute solar data from Kellogg Biological Station (MJ)
- w - wind (m/s)
- lev - water level above level logger probe (ft)
- f – flow, cubic feet per second (cfs)
- vel - water velocity past velocity probe logger (ft/s)
- t - thermistor for water temperature (ºC)

Database-Coded Instrument List

***'EDT** = Eastern Daylight Time.*
Dates and times are matched on Eastern Daylight Time (EDT). Data queries output 5- and 10-minute interval data from the database, or calculate basic statistics of interest using interval data (e.g., hourly average temperature).

***'r-8ma-sum** = Rainfall gauge #8, at mall (tips/5-min).Site 8.*
This Rainwise® rain gauge is located on northwest corner of the grounds of the Crossroads Mall, Portage, MI. One tip equals 0.01 inches. A rain gauge network was distributed and maintained throughout the drainage and surrounding watershed during the course of the entire study.

***'a-12wlhighrestemp_c** = Air high resolution temperature (ºC) in the wetland. Site 12.*
Hobo air temperature meter mounted in a protective environmental chamber about 1.5 m above wetland surface. Operated 6/99-10/02.

***'a-12 wl%rh** = Air percent relative humidity in the wetland (%). Site 12.*

***'a-12wldewpt** = Air dewpoint in the wetland. Site 12.*

***'a-12wlabshum** = Air absolute humidity (gm/M^3) in the wetland. Site 12.*

***'SlrW_Avg** = Solar radiation average ($watts/m^2$) in the wetland. Site 12.*
This instrument, maintained by K&A, measured incoming solar radiation and showed some dampening in sensitivity during the 2002 study year. Additional, comparable data were included from the Kellogg Biological Station, Michigan State University. K&A pyranometer mounted 2 m above wetland surface. Operated 6/99-10/02.

***'kbs hourly total MJ** = Solar radiation hourly total in MJ, from Kellogg Biological Station, MI.*
Site location is approximately 20 miles NE of Consolidated Drain.

***'w-speed(m/s)** = Wind speed (m/s) in the wetland. Site 12.*

K&A anemometer wind speed and direction mounted 2 m above wetland vegetation. Operated 5/01-10/02.

'lev-25maru = *Level mall runoff (ft). Site 25.*
This site is located in the manhole (MH-9) off the mall parking lot. Level data recorded using Global Water® level logger. In dry conditions, logger is in air. During and after runoff events, logger measures water level in MH-9. Operated 6/02-10/02.

'lev-28seinis = *Sediment In Level (ft). Site 28.*
An ISCO area velocity meter measured both level of the sediment forebay and velocity of water entering the forebay through the 42" pipe. Operated 7/02-10/02.

'lev-20wpou = *Level Wetpond Out (ft). Site 20.*
Wetpond level was recorded with a Global Water® level logger mounted on the outflow structure. Operated 8/01-10/02.

'lev-19wlou = *Level Wetland Out (ft). Site 19.*
Level was measured at the bottom notch of a 3-stage outflow restriction weir. Operated 7/02-10/02.

'lev-23sc = *Level Schuring Drain (ft). Site 23.*
Level in Schuring drain was recorded using a Global Water® level logger mounted in the channel. Operated 7/02-10/02.

'lev-03drdopost = *Level Consolidated Drain Down Post –Construction (ft). Site 03.*
An ISCO level logger measured level in the Consolidated Drain channel. Characteristics of stormwater at this site (volume, temperature) were a combination of all upstream inputs from all three subwatersheds (Schuring, Mall, Consolidated Drain). Operated 8/99-10/02.

'f-25maru = *Mall runoff flow (cfs). Site 25.*
MH-9 level readings converted to flow. This flow calculation was not a simple direct level to flow translation. There is very little elevation drop between the mall and the sediment forebay. Larger runoff events cause water to back-up in the system and eventually overflows weir #1 and weir #2 which spill over into the new open Consolidated Drain channel. As runoff water moves through the treatment system, water elevation drops at a similar rate in the pipes (seen at the mall MH-9), forebay, and wetpond. This is a pipe storage condition.

'f-25maruwe2 = *Weir 2 overflow into open drain channel (cfs). Site 25.*
Level MH-9 was used to calculate overflow at Weir #2 into the open Consolidated Drain channel.

'f-25maruwe1 = *Weir 1 overflow into open drain channel (cfs). Site 10.*
Level MH-9 was used to calculate overflow at Weir #1 into the open Consolidated Drain channel.

'f-28seinis - *Sediment In Flow (cfs). Site 28.*

An ISCO area velocity meter measured level of the forebay and velocity of runoff entering the forebay through the 42" pipe. Low flow values were not accurate but event related high flow values have been verified.

***'f-20wpou** = Flow Wetpond Out (cfs). Site 20.*
Wetpond outflow was calculated using pond elevation and outflow structure geometry.

***'f-20wpouover** = Flow Wetpond Out Overflow (cfs). Site 20.*
Wetpond emergency spillway overflow was calculated based on pond elevation. At no time during 2002 did the wetpond overflow through the emergency spillway.

***'f-19wlou** = Flow Wetland Out (cfs). Site 19.*
Flow was calculated using water elevation and weir stage/flow characteristics.

***'f-23sc** = Flow Schuring Drain (cfs). Site 23.*
Flow was calculated from a stage-discharge relationship developed for the Schuring Drain channel.

***'f-03drdopost** = Flow Consolidated Drain Down Post–Construction (cfs). Site 03.*
Flow was calculated using a stage-discharge relationship developed at this location during 2002.

***'vel-28seinis** – Sediment In Velocity (ft/s). Site 28.*
Velocity of water entering the sediment forebay through the 42" pipe. Velocity was accurate and was independently confirmed during high flow. However, low flow conditions were not accurately represented, such as 1) during dry weather periods, and 2) when pipe-stored water gradually drained through the 42" pipe to the forebay following large runoff events.

***'t-25maru.1** = #1 Thermistor Mall Runoff (°C). Site 25.*
This thermistor was located in the manhole (MH-9) just off the mall parking lot. In this manhole, runoff from all portions of the mall combines before entering Weir #2. In no flow conditions, this thermistor is typically resting on or in the sediments and standing water at the bottom of the structure. In low flow and high flow conditions, this thermistor was inundated. Operated 7/01-10/02.

***'t-26maru.2** = #2 Thermistor Mall Runoff (°C). Site 26 (same location as site 25).*
This thermistor was located in the manhole (MH-9) just off the mall parking lot. The thermistor is suspended from the manhole grate, and hung just above of the bottom of the concrete structure. In no flow conditions, this thermistor was dry and suspended about 1' off of the bottom of the concrete structure, and its daily temperature fluctuations likely track air temperature in the underground structure. During very low flow, this thermistor may not be inundated. During mid to high flow, this thermistor was inundated and was likely the most accurate of the two mall thermistors in measuring peak stormwater temperature during the event. Operated 6/02-10/02.

***'t-10he** = Thermistor Headwall (°C). Site 10, Weir #1.*
This thermistor was suspended inside of Weir #1 and measured runoff temperature from the basin upstream of the mall parking lot. During dry conditions, this thermistor may be dry or

inundated in standing water. During low and high flow conditions, this thermistor was inundated. Operated 4/01-10/02.

't-16riou = *Thermistor Rink Out (°C). Site 16.*
Temperature of the groundwater-fed fountain/rink overflow was measured. The fountain flowed during late spring, summer, and early fall during daylight hours. Baseflow was set at about 1 cfs. When the fountain was shut down or undergoing service/cleaning, no overflow wetted the thermistor and it measured air temperature. Operated summers 6/01-10/02.

't-28seinpi =*Thermistor Sediment In Pipe (°C). Site 28.*
This thermistor was located at the bottom of the mouth of the 42" conveyance pipe at the sediment forebay-in location. During runoff events, this thermistor measured runoff temperature entering the forebay (combined flow from the upstream watershed and mall). Following large events, pipe stored water slowly drained into the forebay for hours to days following the event and the readings represented the temperature of that pipe stored water. During dry periods, there was no measurable flow into the forebay (though we suspect some sustained inflow given the full condition of the forebay), and the temperature information reflected the temperature at the bottom of the pond. Operated 7/02-10/02.

't-21sein = *Thermistor Sediment In (°C). Site 21.*
This thermistor was deployed out of the flow stream but near the 42" conveyance pipe. It was mounted mid-water column and always inundated. Readings from this unit best represented the pond temperature as a whole, particularly useful to characterize the thermal condition of the pond before storms, after storms, and during diurnal changes. Operated 7/01-10/02.

't-22seou = *Thermistor Sediment Out (°C). Site 22.*
This thermistor was mounted in one of the three sediment outflow pipes. The unit sat in the bottom of the pipe. During wet weather and at times of elevated volume of the forebay, this thermistor was inundated and readings represented outflow temperature. During dry weather periods, this thermistor was in a shaded pipe, but unpredictably inundated, and therefore, readings did not represent outflow temperature. Operated 7/01-10/02.

't-20wpou = *Thermistor Wetpond Out (°C). Site 20.*
This thermistor was mounted on the outflow structure, near the flow stream, and was always inundated. Readings from this thermistor represented: 1) outflow temperature during wet weather and outflow; 2) and pond temperature during dry weather. Operated 7/01-10/02.

't-12wlmi1a = *Thermistor Wetland Mid 1 a (°C). Site 12.*
This thermistor was mounted in a perforated standpipe. During dry weather, this thermistor was exposed to air temperature when the wetland became sufficiently dry. When the wetland was filling, full, and draining post-runoff events, the readings represented upper wetland temperature in a grassy/cattail area. Operated 6/01-10/02.

't-19wlou = *Thermistor Wetland Out (°C). Site 19.*
This thermistor was mounted in an anchored perforated pipe in the outflow stream. During dry weather, this thermistor was exposed to air temperature when the wetland became sufficiently dry and outflow stopped. When the wetland out was flowing the readings represented wetland

outflow temperature at the end of the treatment system, just prior to joining the open Consolidated Drain channel. Operated 7/01-10/02.

't-23sc = *Thermistor Schuring Drain (°C). Site 23.*
This thermistor was mounted in a standpipe in the flowstream of the Schuring Drain Channel. Baseflow in this channel is primarily groundwater fed, but untreated stormwater runoff from the Schuring subwatershed and adjacent city drain pass through this channel and enter the consolidated drain channel. Operated 7/01-10/02.

't-27drabwl = *Thermistor Drain Above Wetland (°C). Site 27.*
This thermistor was mounted in a standpipe in the flowstream of the consolidated drain channel just upstream of the Schuring channel confluence. Water temperature here was influenced by, rink outflow/baseflow volume and temperature, as well as additional thermal inputs from groundwater, minor untreated stormwater inputs, radiation, and air temperature. Operated 5/02-10/02.

't-03drga = *Thermistor Consolidated Drain near Garden Lane Ave. Site 03.*
This thermistor was mounted in a standpipe in the Consolidate Drain channel in the streamflow and was never exposed to air. Downstream of this site, flow enters Portage Creek. Characteristics of stormwater here (volume, temperature) are a combination of all components including the Schuring and Consolidated Drain subwatersheds. Operated 6/99-10/02.

't-04drgaup = *Thermistor Consolidated Drain upstream of Garden Lane Ave. Site 04.*
This thermistor was mounted in a standpipe in Portage Creek about 20 m upstream of the confluence of Portage Creek and the Consolidated Drain. Operated 6/99-10/02.

't-05drgaup = *Thermistor Consolidated Drain downstream of Garden Lane Ave. Site 05.*
This thermistor was mounted in a standpipe in Portage Creek about 50 m downstream of the confluence of Portage Creek and the Consolidated Drain. Operated 6/99-10/02.

't-06drgaup = *Thermistor Consolidated Drain upstream of Garden Lane Ave. Site 06.*
This thermistor was mounted in a standpipe in Portage Creek about 1000 m upstream of the confluence of Portage Creek and the Consolidated Drain. Operated 6/99-10/02.

APPENDIX C

BIOASSESSMENT

C.1 Bioassessment, Multiple Stressors, Urban Runoff, Site Characteristics, and the Isolation of Thermal Effects

Urbanization causes numerous stressors to aquatic ecosystem health (Paul and Meyer, 2001). Holistic water quality monitoring involves the increasing usage of multimetric biotic indices to track long-term aquatic health. These indices are not designed to determine short-term impacts associated with individual stressors (e.g., temperature).

Outside mesocosm level community manipulations (Swanson et al., 2001), there are few places ideally suited (Consolidated Drain included) to isolating thermal impacts on the biota from multiple stressors. The Consolidated Drain was formerly recognized as the single largest contributor of sediment and pollutants to Portage Creek. Construction of the treatment system has changed the dynamics of sediment, nutrient, and other pollutant loadings as well as the hydrologic and thermal regime (Appendix A). The study portion of Portage Creek and most of the downstream reaches are surrounded by impervious area that greatly exceeds 10-15% of the drainage, a level generally recognized to negatively impact biotic integrity in receiving streams with or without BMPs. Thus, isolating a short-term, cause-and-effect relationship of community recovery due to the altered thermal regime was not expected.

Isolating thermal impacts associated with this study is further complicated by the input from Schuring Drain, as well as the potential for occasional treatment system overflows during large storms, passing relatively unmitigated runoff through the drain channel and directly into the creek. In addition, during spring and summer months, the drain channel flow is supplemented by a cool water baseflow from a groundwater-fed fountain/ice rink facility upstream of the treatment system. This flow passes through the open drain channel with little shading and currently experiences large daily temperature fluctuation due to exposure to solar radiation. As vegetation in and around the ponds and channels recovers, the treatment system components will gain more shading with each successive growing season, and it is expected that diel temperature variation will subside.

Despite these complications, there is value in the biotic sampling for our understanding of the overall effect of the treatment system; however, cause and effect relationships between the change of thermal regime and community response were not attempted. The initial characterization of the system is appropriate for long-term monitoring and provides an opportunity for follow-up monitoring.

C.2 Application of a Regionally-Specific Stream Thermal Regime Classification to Assess Fish Community Structure

Traditional laboratory determination of aquatic temperature limits of stream biota focuses on the effects of maximum or average temperature. However, it has been suggested more recently that the range of diel variation is important and that human activity can influence such diel variations in streams (Matthews, 1987; Wehrly, 1999; Schrank et al., 2003; Johnstone and Rahel, 2003).

Wehrly et al. (2003) developed a regionally specific relationship between stream thermal regimes and fish community structure in the Lower Peninsula of Michigan based on regimes of 670 monitored or modeled sites of which 370 contained species richness information. By identifying mean temperature categories (cold, <19°C; cool, 19 to <22°C; and warm, ≥22°C), and three temperature fluctuation categories (stable, <5°C; moderate, 5 to <10°C; and extreme, ≥10°C) a 3 × 3 matrix was developed with nine stream thermal regime categories. Species specific fish presence and abundance data were evaluated against this thermal classification system to develop thermal regime preference and suitability ratings.

The relationship was developed using stream temperature during the first three weeks in July. Weekly max and mean temperatures were recorded and used to produce an average mean and max for the time period. July average stream temperature was calculated by averaging the minimum and maximum measurements and was found to be within 1°C of continuously measured comparison sites. July weekly temperature fluctuation was calculated as the difference between the minimum and maximum. This tool is the first to use both average stream temperature and average weekly fluctuation to characterize fish communities based on species-specific thermal criteria.

The thermal regimes of Michigan streams are highly influenced by basin and local scale attributes (Wehrly et al., 1998). Stream thermal regime changes associated with localized thermal enrichment of urban runoff can be both acute (increasing temperature fluctuation) and chronic (increasing average temperature). Consequently, changes of fish community structure can be linked to the localized effects of thermally enriched urban runoff.

C.3 Methods

Bioassessments conducted for the project followed the State of Michigan's Qualitative Biological and Habitat Survey Protocols for Wadeable Streams and Rivers (MDEQ, 2002) adapted from the U.S. EPA's Rapid Bioassessment Protocols for Use in Streams and Rivers (U.S. EPA, 1999). As opposed to the use of the more traditional water quality analyses for determining the quality of a river system, bioassessments allow for the integration of chemical, physical and biological factors into ecosystem health analyses.

Each sampling location in this study consisted of a 100-meter long reach of stream (or drain), which was identified during the first bioassessment field visit and was subsequently used for each sampling event thereafter (Figure C-1).

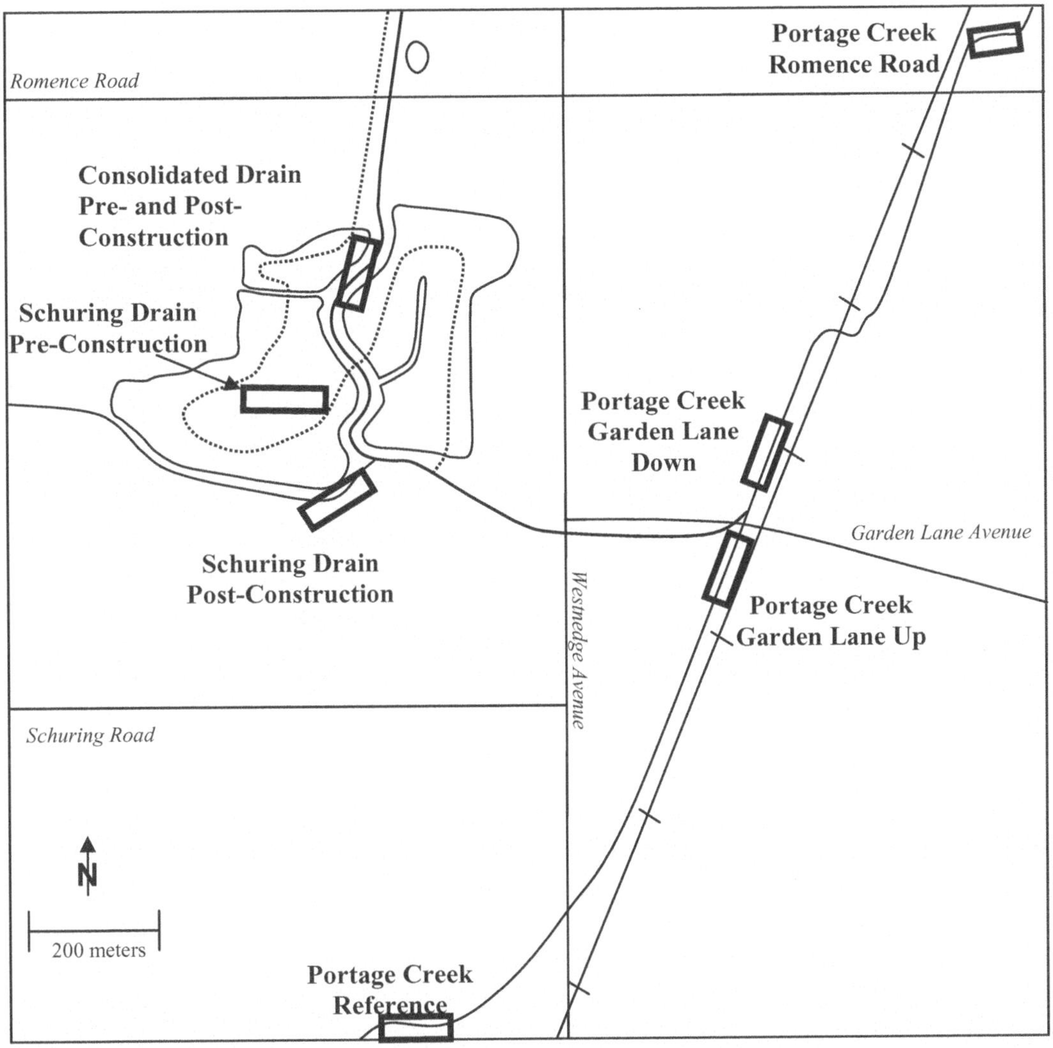

Figure C-1. Bioassessment sampling sites pre- and post-construction of the Consolidated Drain stormwater treatment system, Portage, Michigan. Sample sites are located in the Schuring Drain channel, Consolidated Drain channel, and Portage Creek receiving stream (N=4).

The six sampling locations included:

- Consolidated Drain – pre-construction;
- Consolidated Drain – post-construction;
- Schuring Drain – pre-construction;
- Schuring Drain – post-construction;
- Portage Creek Romence Road - downstream of drain confluence (at Romence Road);
- Portage Creek Garden Lane Down - downstream of drain confluence (at Garden Lane Ave.);
- Portage Creek Garden Lane Up - upstream of drain confluence (at Garden Lane Ave.);
- Portage Creek Reference - upstream reference site.

The high quality reference site identified for this study is a section of Portage Creek located 1000 meters upstream of the Consolidated Drain confluence. This site has had recent trout habitat improvements and represents a base case condition for Portage Creek despite some recognized impacts from urbanization. The upstream reference site was sampled in duplicate for QA/QC purposes.

The intent of this bioassessment study was to compare each site to itself over the course of the study, rather than to compare each site to a regional or study-specific reference site, which is the common practice with a bioassessment. This alternative approach was chosen because the study sites were not homogeneous and a true reference site application would not yield meaningful ratings. In addition, chosen bioassessment survey methods are not currently used or designed for statistical inference but instead provide a simple basis of comparison based on straightforward count and diversity data typical for this stream type. Standardized bioassessment methodology was established and followed in the study to provide an opportunity for follow-up study once the treatment system has undergone multiple growing seasons and vegetation is re-established.

The bioassessment portion of this study consisted of three components including fish, benthic macroinvertebrates, and periphyton. A fish survey was conducted on December 20, 2001. Benthic macroinvertebrate surveys were conducted on: May 25, 2000; October 2, 2000; May 14, 2001; October 22, 2001; and May 28, 2002. Periphyton surveys were completed on: May 22, 2000; August 31, 2000; November 6, 2000; June 11, 2001; August 13, 2001; November 29, 2001; and May 23, 2002.

C.3.1 Fish

Fish collections were conducted on December 20, 2001 at six locations sampled previously for benthic macroinvertebrates and periphyton. Fish were collected using a Smith-Root® backpack electrofisher producing pulsed, direct current. Transects were sampled in an upstream direction for a maximum of 45 minutes or until the entire transect had been covered; block nets were not used given the relative inactivity of most fish during December. Fish were identified on-site and returned to the stream. One live individual was brought to the lab for identification and then returned to the stream.

C.3.2 Benthic Macroinvertebrates

A composite sample was collected for each sampling location. A D-frame dip net with 500 μm mesh was used to sample 20 jabs within each 100-meter long reach. The jabs were collected from habitats (cobble, vegetation, snags, sand, etc.) in the proportion represented. The net was emptied into a white collection pan on shore where organisms were placed into a labeled plastic container using forceps or an eye-dropper, depending on organism size. The net was rinsed with deionized water between sampling locations. Benthic macroinvertebrate samples were collected in May and October of 2000 and 2001. A May 2002 sample was also collected. The samples were preserved with 95% ethanol and refrigerated. All samples were submitted to Phycotech, Inc. of St. Joseph, MI for identification to species level.

C.3.3 Periphyton

The same six bioassessment sites were sampled for periphyton. A composite sample was collected for each sampling location. Twenty samples were collected from substrate (rocks, submerged vegetation, submerged woody vegetation, sand, etc.) in the proportion represented within each reach. The samples were collected using a combination of eye dropper and toothbrush for scraping. Deionized water was used to rinse the sample into a labeled plastic container. Periphyton were collected in May, August and November 2000. Samples were also collected in 2000 in June, August and October. The spring 2001 sampling event was delayed until June, due to numerous wet weather events in May 2001 that did not allow for the monitoring protocol-specified recovery periods for the periphyton community prior to sampling. The samples were preserved with gluteraldehyde, wrapped in foil to inhibit light, and refrigerated. All samples were submitted to Phycotech, Inc., of St. Joseph, MI, for identification to species level.

C.4 Results and Discussion

Habitat was evaluated during the spring sampling each year during the study, with the exceptions of the Consolidated Drain and the Schuring Drain, which were under construction during Spring 2001. The RBP scores for each site over the three-year study period are presented in Table C-1.

Table C-1. U.S. EPA Rapid Bioassessment Protocol habitat scores for Portage Creek (PC, four locations), Consolidated Drain, and Schuring Drain during 2000-2002.

	Score		
Site	**2000**	**2001**	**2002**
PC Romence	156	169	150
PC Garden Down	111	130	96
PC Garden Up	100	102	90
PC Reference	143	140	125
Consolidated Drain	124	NS	118
Schuring Drain	116	NS	96

NS = Not sampled

The two drain locations represent highly disturbed/altered habitat conditions. The Consolidated Drain and Schuring Drain underwent major modifications during this study. The Consolidated Drain was reconfigured during construction and has been heavily planted with riparian vegetation which experienced its first full growing season in 2002. The Schuring Drain was changed from a ponded water system to a flowing groundwater fed channel.

The Portage Creek sites are quite variable. The upstream "Reference" site has been highly modified within the last ten years to create favorable trout habitat structure and a deep, narrow channel. The aquatic habitat at this site is good with large cobble/rock substrate and higher flow velocities, but it is considerably different from unmodified Portage Creek habitat that consists of sandy substrate and lower flow conditions common in this reagion. This "Reference" site functions as an indicator of natural seasonal fluctuations in the sampled populations during this study.

The Garden Lane Up sampling site (Figure C-1) is located just upstream of the Consolidated Drain confluence. The aquatic habitat here is poor with little heterogeneity. This site, too, serves as a reference site to compare population fluctuations noted throughout the study in the downstream sites. This Garden Lane Up survey site is most comparable to the Garden

Lane Down site which also has relatively poor aquatic habitat with little heterogeneity (Table C-1).

The Romence Road site is the furthest downstream sampling site on Portage Creek. The in-stream habitat is heterogeneous with both runs and plunge pools present with submerged woody debris and considerable in-stream submerged vegetation. The riparian zone is well-vegetated and provides a canopy over much of this reach. This site is most directly comparable to the Reference site (Table C-1).

C.4.1 Thermal Characteristics and Classification

Like many coldwater streams in Michigan, Portage Creek gains considerable baseflow from groundwater input (Wehrly et al., 2003), effectively cooling the stream during the summer months between the headwaters to the confluence with the Consolidated Drain. This decline in temperature is slight but detectable (Kieser and Associates unpublished data), and it is the primary reason that the Michigan Department of Natural Resources (MDNR) continues to stock this portion of the stream with brown trout (Jay Wesley, MDNR personal communication).

Stream temperature information was collected at Garden Lane Down, Garden Lane Up, and Reference sites. Using the Wehrly et al. (2003) thermal classification, Portage Creek is considered "cool" (average weekly temperature 19.2-19.4°C) and "moderate" (fluctuation 6.9-8.0°C) for both pre- and post-construction upstream and downstream of the Consolidated Drain Confluence.

Temperature information collected in the Consolidated Drain, just upstream of its confluence with Portage Creek indicates that 1999 pre-construction input from the drain would be classified as "cold" (average temperature 18.2°C) and "moderate" (fluctuation 8.6°C). Post-construction 2002 input is classified as "cold" (average temperature 18.9°C) and "extreme" (fluctuation 10.4°C).

C.4.2 Fish

Table C-2 lists species captured during the post-construction fish sampling at all six sites. Fish capture data match quite well with patterns for coolwater-moderate stream fish communities. The community is dominated by cold and coolwater guild species. The presence of other less dominant warmwater species is to be expected given Portage Creek's connections to headwater wetland complexes, drainage systems, and a downstream impoundment. The species assemblage resembles MDNR brown trout stock assessment sampling conducted in 1997 (MDNR unpublished data).

Table C-2. Fish species collected during backpack electrofishing efforts on December 21, 2001 (post-construction).

Species	Common Name	Sampling Locations (*Sampling effort in minutes*) Reference Up (30)	Garden Up (*27*)	Garden Down (*20*)	Romence Down (*35*)	Consolidated Drain (*30*)	Schuring Drain (*25*)
Rhinichthys cataractae	Blacknose Dace	8	2	33	21	-	-
Lepomis macrochirus	Bluegill	-	-	1	-	-	2
Culaea inconstans	Brook Stickleback	-	-	2	-	-	-
Salmo trutta	Brown Trout	1	-	-	1	-	-
Umbra limi	Central Mudminnow	-	1	1	-	-	-
Semotilus atromaculatus	Creek Chub	6	1	7	3	-	14
Carassius auratus	Goldfish	-	1	-	-	-	-
Lempomis cyanellus	Green Sunfish	1	-	2	1	-	-
Etheostoma nigrum	Johnny Darter	-	2	5	12	-	-
Ichthyomyzon spp.	Lamprey (Family)	-	-	1	1	-	-
Micropterus salmoides	Largemouth Bass	-	-	-	-	-	-
Esox americanus	Grass Pickerel	-	1	-	-	-	-
Cottus spp.	Sculpin (Family)	48	48	36	17	-	-
Lepomis spp.	Sunfish (Family)	-	-	1	-	1	26
Gasterosteus aculeatus	Threespine Stickleback	-	-	-	-	-	-
Catostomus commersoni	White Sucker	-	-	3	1	-	2

The brown trout population in Portage Creek is maintained through annual stocking (Jay Wesley, MDNR, personal communication). Multiple year classes of stocked trout persist both up- and downstream of the Consolidated Drain confluence and trout have responded well to habitat improvements at the Reference site (MDNR unpublished data). The 1997 assessment survey did suggest that one individual was believed to be the result of natural reproduction, indicating the potential of Portage Creek to support natural reproduction.

Only one individual fish was captured in the Consolidated Drain channel (Table C-2). Baseflow in the channel was maintained by minimal groundwater inputs during the time period sampled and was less than 0.028 m^3/s in an exposed channel with little fish habitat and less than 5" deep. The Consolidated Drain may have provided adequate fish habitat during the warmer seasons when baseflow is supplemented by outflow from a groundwater fed fountain at about 0.028 m^3/s. However, the Consolidated Drain was not designed specifically for year round fish habitat and is subject to winter ice cover. As vegetation continues to re-establish a canopy in and around the channel, temperature extremes during the warm seasons and habitat structure will improve.

Schuring Drain was the only site co-dominated by sunfishes (26 of which were < 1", likely young-of-year; Table C-2). It also lacked species common to the stream sites (e.g., blacknose dace and sculpin). The co-dominance of small sunfishes may be related to intermittent connections to headwater wetlands or nearby ponds and open standing water channels associated with nearby greenhouse activity. In addition, Schuring Drain changed from a ponded ditch to a relatively deep, narrow groundwater fed channel that is now subject to flashy stormwater flows from the Schuring subwatershed. This likely further explains the co-dominance of creek chub, a species whose thermal category preferences is coldwater-extreme.

Figure C-2 compares the fish species percent occurrences up- and downstream of the drain confluence at the upstream Reference site and downstream Romence Road site during pre-construction (MDEQ unpublished data, 1997) and post-construction (2001). Fish species maximum weekly average temperature tolerances are listed in increasing order from left to right (Eaton and Scheller, 1996). There are several points to note in this plot.

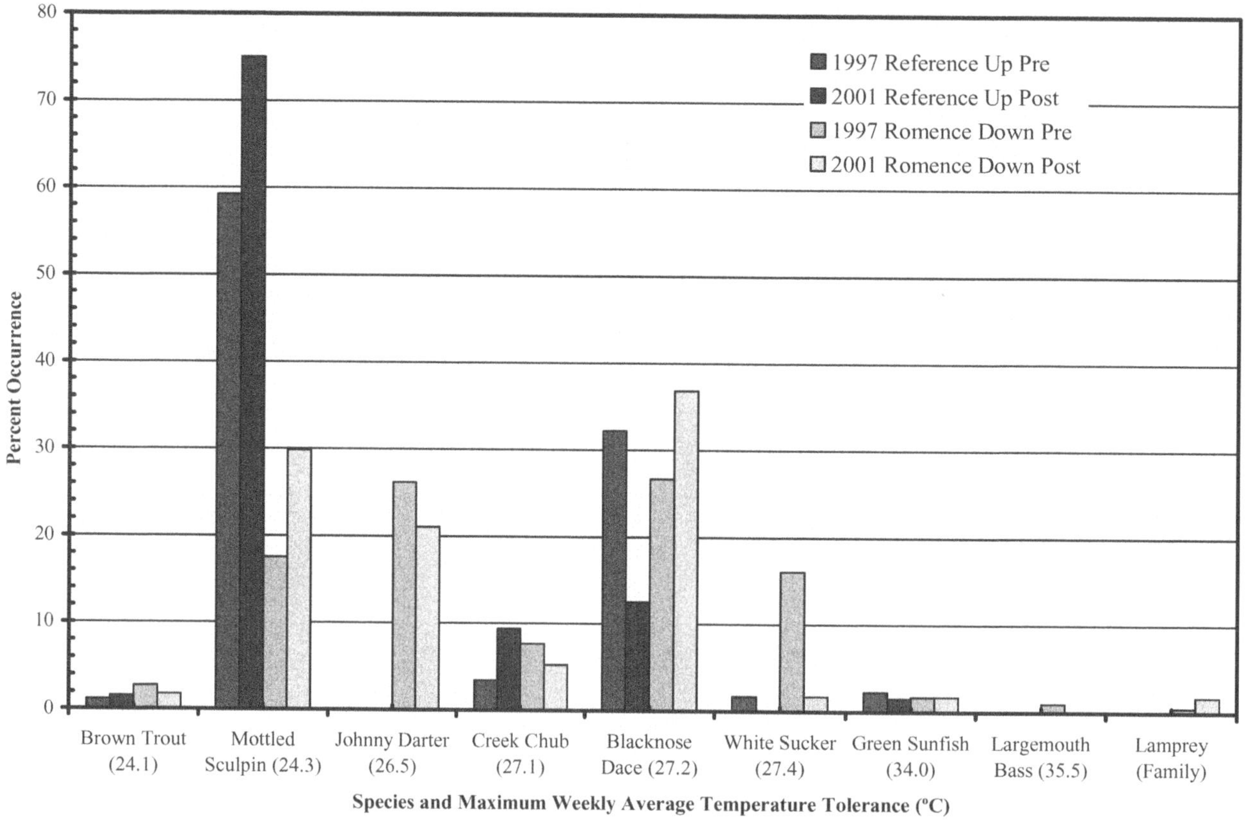

Figure C-2. Fish species percent occurrence sampled upstream and downstream of the Consolidated Drain at the reference site (upstream) and Romence (downstream) pre-construction (1997) and post-construction (2002).

First, the most thermally restricted species captured, brown trout, occurred at the up- and downstream sites both pre- and post-construction of the stormwater treatment system, making up 1-3% of species captured (Figure C-2). Second, the upstream Reference site was dominated by mottled sculpin, a species that finds cold-stable temperature regimes optimal and cold-moderate suitable. However, mottled sculpin are not likely to occur in streams with more extreme thermal fluctuation. Blacknose dace were the next most common species at this location. They are more thermally tolerant preferring cold-stable regimes and finding cold-extreme and cool-moderate regimes suitable. Third, at the downstream Romence site dominance shifted to blacknose dace with a more equal contribution from mottled sculpin, as well as a mix of other more tolerant species including johnny darter, creek chub, and white sucker (Figure C-2). Fourth, although it offers no conclusive evidence, pre-and post-construction fish assemblage structure at the downstream Romence site may be shifting toward that seen at the upstream Reference site with a shared dominance of mottled sculpin and blacknose dace. As the Consolidated Drain treatment system matures, and as shown in previous chapters, average temperature and fluctuation at the downstream sites was expected to decline. Continued change may play a role in further fish assemblage structure shifting.

Figure C-3 compares percent occurrence of fish post-construction at the upstream site (Garden Lane up) and downstream site (Garden Lane down). These sites have very similar habitat to one another, with lower quality, more homogenous habitat compared to the Reference and the further downstream Romence sites (Figure C-1). The first observation (from Figure C-3) is the lack of trout captured at these sites, most likely due to a lack of structure and cover typically utilized by trout during the early winter (time of year sampled). Secondly, species composition is similar to that described in Figure C-2, with the upstream site dominated by mottled sculpin and the downstream site having a more balanced community with a greater contribution of species more tolerant of higher average temperatures and diel temperature fluctuations.

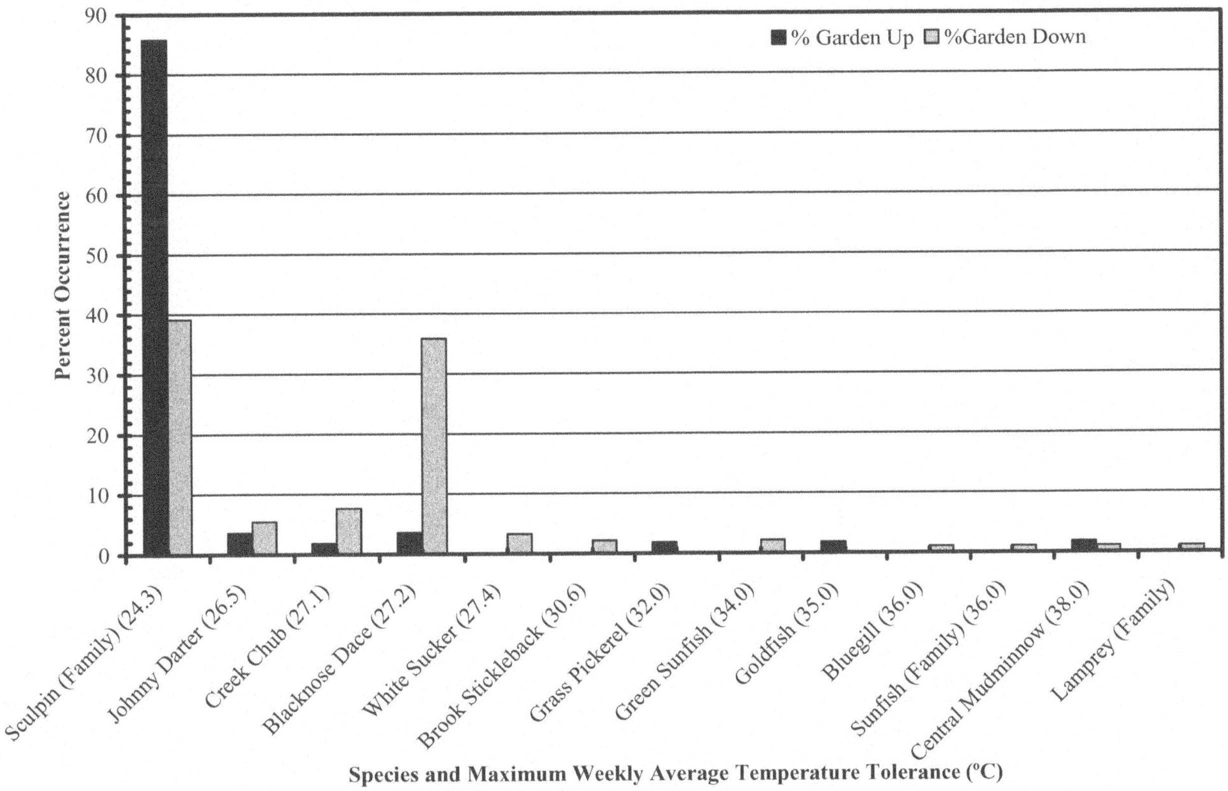

Figure C-3. Fish species percent occurrence sampled in Portage Creek immediately upstream (Garden Up) and downstream (Garden Down) of the Consolidated Drain during winter 2002.

C.4.3 Macroinvertebrates

The benthic macroinvertebrate community was sampled from each survey location reach five times during this study. Exceptions to this are the Consolidated Drain and Schuring Drain which were not sampled during Spring 2001 due to construction and alteration of the drain channels.

Nine metrics were used to evaluate the six sampling reaches. The metrics were compared to show change over time at each site. The nine metrics are:

- Metric 1 - Number of Taxa;
- Metric 2 - Number of Mayfly Taxa (Ephemeroptera);
- Metric 3 - Number of Caddisfly Taxa (Trichoptera);
- Metric 4 - Number of Stonefly Taxa (Plecoptera);
- Metric 5 - % Mayfly Composition;
- Metric 6 - % Caddisfly Composition;
- Metric 7 - % Contribution of the Dominant Taxa;
- Metric 8 - % Isopods, Snails and Leeches;
- Metric 9 - % Surface Dependent Organisms.

The chosen metrics are those used by the Michigan Department of Environmental Quality Procedure 51 (MDEQ, 2002).

The data have been grouped into a pre-construction period (May 2000, October 2000 and May 2001) and a post-construction period (October 2001 and May 2002). The Reference site was sampled in duplicate as a Quality Assurance measure. Results from this site are designated Ref #1 and Ref #2 and represent the duplicate samples. The May 2002 results are questionable due to the small number of organisms reported in each sample and are believed to be the result of contract lab error. The results were averaged with the other sample to give another set of post-construction data. The results were consistently low across all sampling sites, so the error was evenly distributed across all sites.

No stonefly taxa were found at any of the survey sites during this study, though stoneflies are typical components of coldwater streams. This group is the least tolerant of thermal stress with temperature tolerances generally below 20ºC (Galli, 1990)

All benthic macroinvertebrate metric results are shown in Figure C-4. In the Consolidated Drain (Figure C-4a) the changes to note are that while the number of taxa (Metric 1) present from the pre-construction to post-construction periods did decrease at this site (from 13.5 to 8), caddisfly species (Metric 3) were present after construction and the percent contribution of the dominant taxa (Metric 7) greatly decreased in the post-construction period (from 52.5 to 13.5%) showing greater community diversity. Also, the percent contribution of Isopods, Snails and Leeches (Metric 8) decreased. High percentages of these organisms are good evidence of stream degradation (MDEQ, 2002). Additionally, the percent composition of Surface Dependent Organisms (Metric 9) also showed a decrease from 10.5 to 1.5% in the post-construction surveys. High percentages of these organisms can indicate large dissolved oxygen shifts, elevated temperatures and low or erratic flows. This improvement is likely related to physical changes made to the channel, the addition of a supplemental baseflow, and more importantly, the hydraulic mitigation introduced by the stormwater treatment system. The sample area changed from a standing water storm channel to a predominantly stable, lotic habitat with additional manmade structure (e.g., log deflectors, cobble, and constructed riffles).

The Schuring Drain in its pre- vs. post-construction period did not resemble the Consolidated Drain based on the benthic macroinvertebrate metrics (Figure C-4b). No Ephemeroptera, Trichoptera or Plecoptera species were present in either the pre- or post-construction surveys. Additionally, Metric 7 (% Contribution of the Dominant Taxa) increased from 42.5 to 79%, although Metric 9 (% Surface Dependent Organisms) decreased from 30.5 to 1.5% in the post-construction period. These changes likely reflect the drastic channel morphology modification that the Schuring Drain underwent altering it from a ponded-ditch to a channel with soft marl substrate devoid of woody debris and rock material. Significant growth of submerged aquatic vegetation was observed in the first full growing season following construction and should provide habitat and shading with each successive growing season.

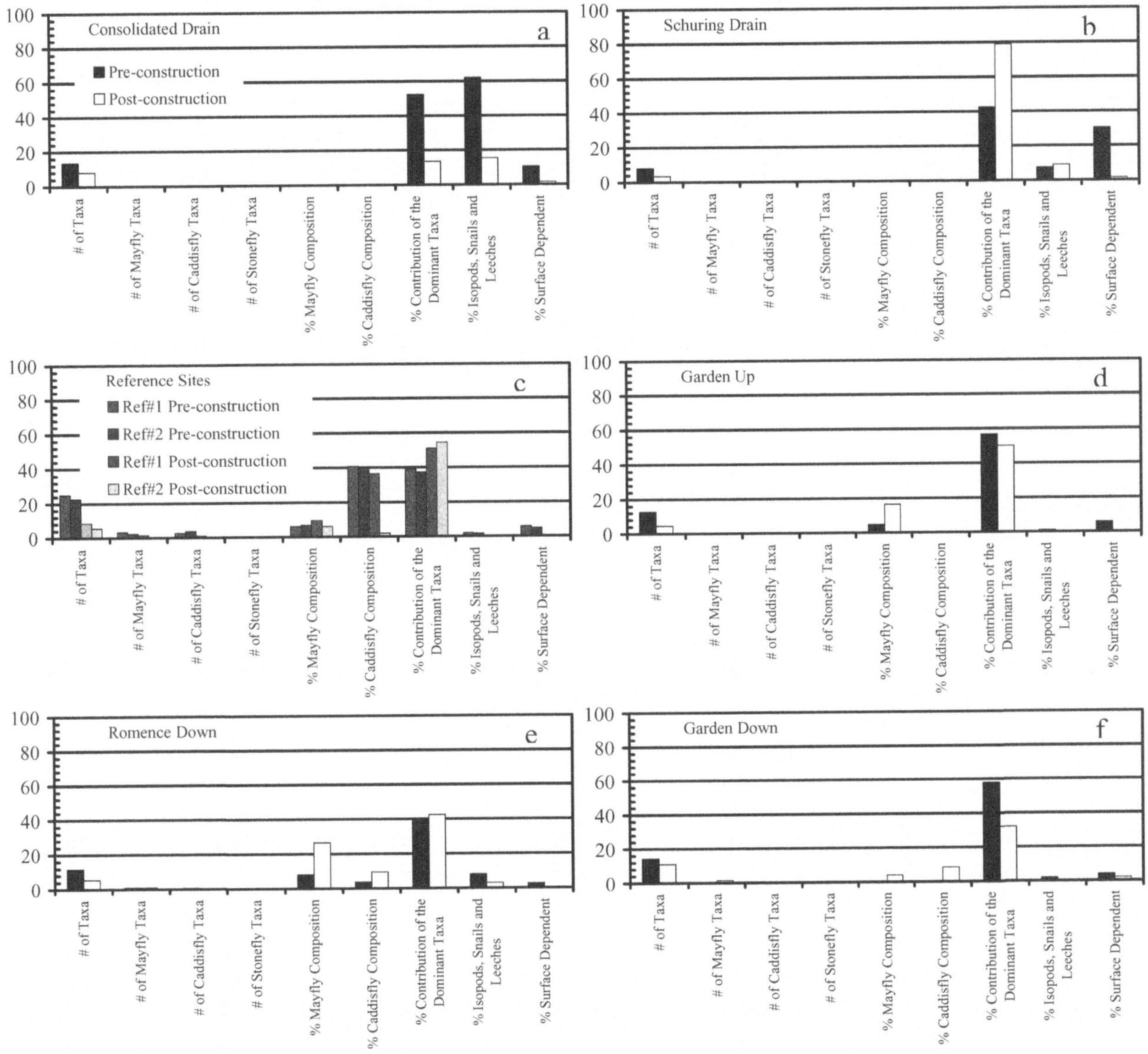

Figure C-4. Benthic macroinvertebrate bioassessment categories. Reference site was sampled in duplicate and labeled Ref#1 and Ref#2.

The Garden Lane Down data (Figure C-4d) show increases in the number of caddisfly species (from 0.3 to 1.5%) post-construction and also decreases in Metrics 7 (58 to 32%), 8 (2 to 0%) and 9 (4 to 2%). At the Romence Road location, the Number of Taxa (Metric 1) (from 11.7 to 5.5) identified at this site again showed post-construction decreases, though Metrics 5 and 6 increased (from 8 to 26.5% and from 3.7 to 9.5%, respectively). Metrics 8 and 9 decreased at this site during the post-construction period.

Comparing these downstream sites to the upstream Portage Creek sites (Figures C-4c and C-4d) shows that a similar trend has been noted in these unimpacted upstream sites. As noted in the downstream Portage Creek and both Drain sites, the total number of taxa collected decreased in the post-construction periods, suggesting either natural fluctuations in populations during these survey years or laboratory error in the May 2002 sample. In either case, these data do not show immediate differences in pre vs. post-construction status of the Consolidated Drain stormwater treatment facility. The one exception to this is the Consolidated Drain channel itself which shows a healthier post-construction system based on the metrics evaluated in this study. Additional post-construction surveys conducted after riparian vegetation is fully established would likely yield a more representative picture of any downstream improvement/effects of the treatment system.

C.4.4 Periphyton

The periphyton community was sampled seven times for each of the six sites during this study. The exceptions to this are the Schuring Drain which was not sampled during the May 2001 survey and the Consolidated Drain which was not sampled during the May or August 2001 survey because construction work prevented sampling in the drains at these times.

Periphyton were collected following U.S. EPA Rapid Bioassessment Protocols for multi-habitat sampling. Natural substrates were sampled per the recommendation of Dr. Ann St. Amand of Phycotech Laboratories Inc. of St. Joseph, MI, a contractor on this project. Organisms identified to the species level where possible.

The data have been grouped into a 'pre-construction' period (May 2000, August 2000, November 2000, and June 2001) and a 'post-construction' period (August 2001, November 2001 and May 2002). The Reference site was sampled in duplicate as a Quality Assurance measure. Results from this site are designated Ref #1 and Ref #2 to represent the duplicate samples. The data are grouped by division: Bacillariophyceae (Diatoms), Chrysophyta (Yellow-green algae), Chlorophyta (Green algae), Cryptophyta (Cryptomonads), Cyanophyta (Blue-green bacteria), and Xanthophyta (Yellow-green algae). Per a recommendation from Dr. St. Amand, two groups of non-motile blue-green bacteria and the *Synechococcus* sp. 1 were omitted from this percentage to yield a more representative picture of the periphyton community. These three groups are extremely small and their numbers can overwhelm the sample relative percent concentration. In the following graphs and discussion, these species have been omitted but their presence in the periphyton samples should be noted.

All periphyton results are presented in Figure C-5. The Consolidated Drain periphyton community (Figure C-5a) shifted from Cyanophyta (blue-green bacteria) to Bacillariophyceae (diatoms). In general, the diatoms are representative of mesothermal conditions (15-30°C) and the blue-green bacteria are dominant in euthermal conditions (> 30°C, Galli, 1990). The presence of yellow-green algae (Chrysophyta) typically suggest cooler (<15 °C) water conditions. The Consolidated Drain system had a relative percent concentration of 7.8% Bacillariophyceae and 89.1% Cyanophyta in the pre-construction survey. In contrast, Bacillariophyceae represented 40.8% of the relative concentration and Cyanophyta 51.9% in the post-construction surveys, a change of +33% and -37.2%, respectively.

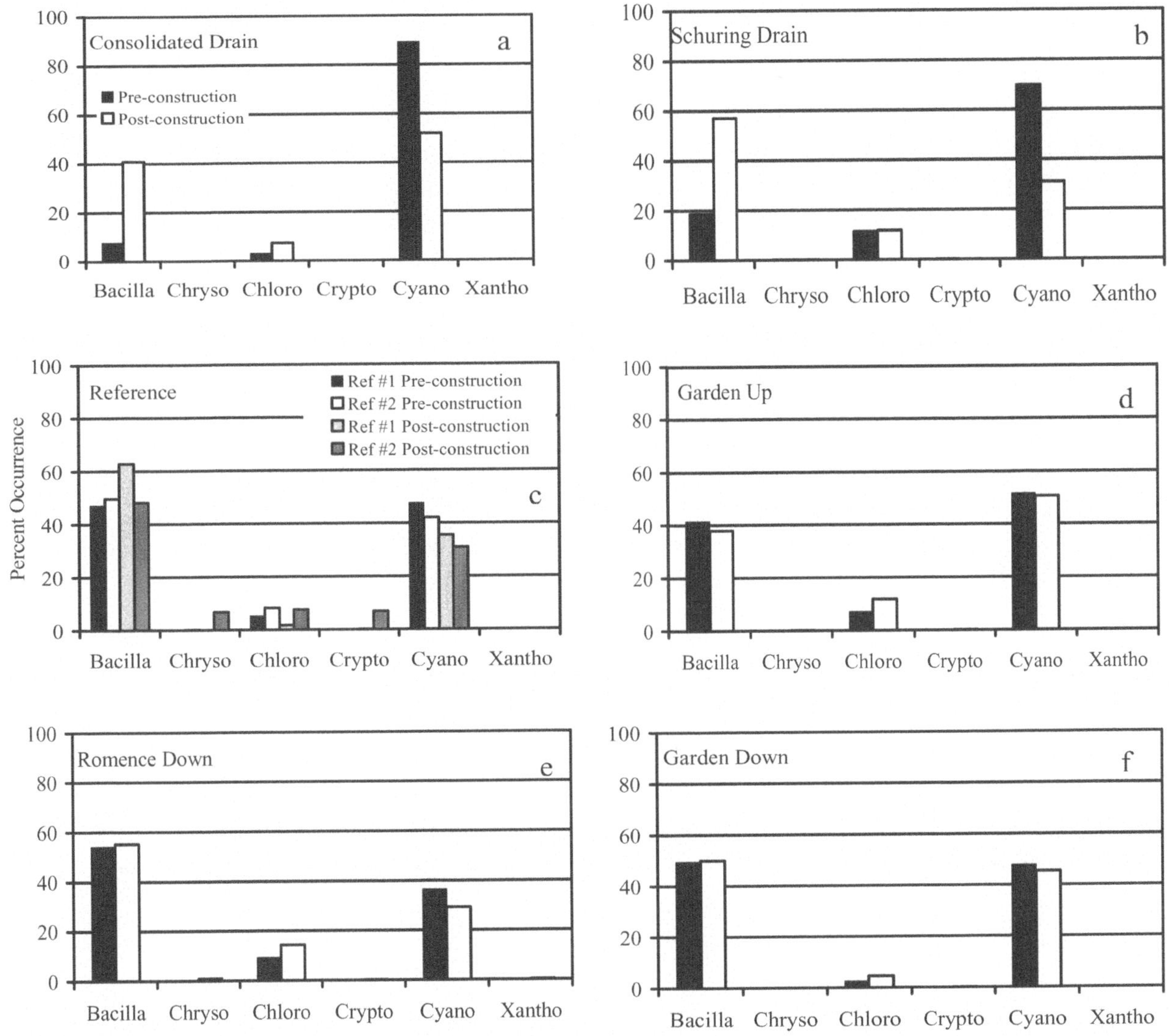

Figure C-5. Percent occurrence of periphyton divisions at six sample sites pre- and post construction of the consolidated drain. Reference site was sampled in duplicate and labeled ref#1 and Ref#2.

The Schuring Drain (Figure C-5b) underwent a similar periphyton community change in the pre-construction vs. post-construction surveys. Bacillariophyceae comprised 18.6% of the periphyton community in the pre-construction period and 57% in the post-construction period. Cyanophyta showed a 38.8% drop in relative percent concentration from pre- to post-construction periods (69.8 to 31%).

The Garden Lane Down and Romence Road periphyton data (Figure C-5e and C-5f) show little difference in the sampled communities with similar average percent concentration values observed for both pre- and post-construction periods. The Garden Lane Up site, located just upstream of the Consolidated Drain confluence on Portage Creek, displayed similar results for the pre- and post-construction survey periods. The upstream Reference site (Figure C-5c) showed little change in the Bacillariophyceae and Cyanophyta percent concentrations during this study period. In one post-construction replicate average sample Chrysophyta and Cryptophyta were detected at concentrations >6% of the relative total. These higher concentrations of these two species were only noted in one of the replicate samples at this site and not at other sites in this study.

Portage Creek sites downstream of the Consolidated Drain did not show immediate obvious changes in the periphyton community from pre- to post-construction during the project study period. The data collected from the Consolidated Drain and Schuring Drain showed a shift in the periphyton community from warm-water tolerant Cyanophyta dominance in the pre-construction period to a cooler-water tolerant Bacillariophyceae dominance in the post-construction period. As noted in the previous section, additional post-construction surveys after the storm water treatment system is fully established would likely yield a more representative picture of the downstream improvements/effects of the treatment system.

C.5 Future

Long-term bioassessment was not an option under this WERF contract, but the project team remains interested in pursuing opportunities to further this work. The establishment of vegetation in the treatment system cells and channels is expected to further enhance its thermal treatment efficiency.

Michigan-specific stream thermal regime and fish community structure relationships were available and provided a species specific conceptual framework on how localized thermal loading may contribute to altered aquatic communities. Directly applicable thermal indices were not available for macroinvertebrates and periphyton and development of such indices was beyond the scope of this project. Complete isolation of thermal impacts on aquatic ecosystems is not realistic, but existing tools can demonstrate the conceptual link between stormwater thermal loads and aquatic ecosystems. Further consideration should be given to stream warming in urbanizing areas as an agent of change in urban coldwater aquatic ecosystems at the watershed scale.

APPENDIX D

MODEL BACKGROUND THEORY

Yi Li
William James

D.1 General Description

In this study, all situations or hydraulic elements involved in an urban stormwater drainage system are generalized into a single description, and applied to four broadly similar flow situations or elements: 1) overland flow; 2) closed conduit or pipe flow; 3) open channel flow; and 4) ponds and wetlands.

Like common energy budget models, our heat budget analysis is based on a basic assumption: the water in each element is completely mixed over every time step, implying that every waterbody has a uniform temperature at any time. In urban drainage, flow in hydraulic elements such as storage in stormwater ponds (wet and dry), flows in pipes and ditches, and overland sheet flow, are usually small and shallow. The characteristics of urban stormwater systems, such as the impact of rain drops on the open water surface, wind shear, high flow velocities in sewers, and rough and discontinuous boundaries, make these flows likely to be completely mixed in a short time. Indeed, the uniform temperature condition has been observed at several shallow reservoirs (Raphael, 1962).

Additionally, minor heat transfer components are ignored, for example exfiltration into streams and ponds, biological or chemical processes within the water body, and internal flow friction. Figure D-1 shows the components used in our general energy budget for a natural open body of water.

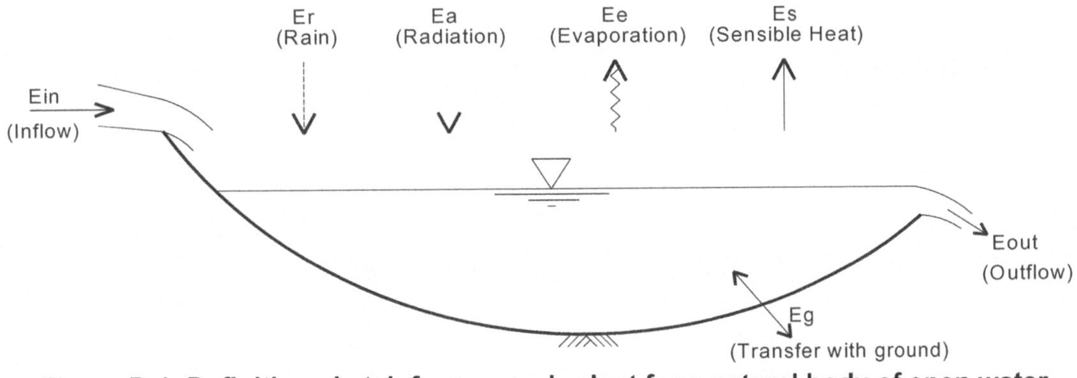

Figure D-1. Definition sketch for energy budget for a natural body of open water.

Taking the waterbody as the control volume, a general energy balance in units of Joules during a given period can be expressed as:

$$\Delta E = Er + Ea - Ee - Es + Ein - Eout + Eg \qquad (1)$$

where: ΔE = total energy increment;
Er = energy introduced by rain;
Ea = net radiation;
Ee = energy loss by evaporation;
Es = energy loss by sensible heat flux;
Ein = energy introduced by inflow;
$Eout$ = energy loss by outflow;
Eg = convection transfer between water and ground.

Note that in the above equation, the ground could be paving, channel bed, pipe material, or pond bed, etc.

From ΔE, we can obtain a new mean temperature of the water body, $Tnew$:

$$Enew = \Delta E + Eold$$

$$\text{or} \quad Vnew \cdot \rho \cdot C \cdot Tnew = \Delta E + Vold \cdot \rho \cdot c \cdot Told \qquad (2)$$

where: $Vnew$ = updated waterbody volume at the end of the period (m^3);
$Vold$ = waterbody volume at beginning of the period (m^3);
ρ = density of water (kg/m^3);
c = specific heat of water $(J / kg \cdot °C)$;
$Told$ = water temperature at beginning of the period $(°C)$;
(ρ and c vary with temperature).

D.2 Heat Transfer Component Details

The above seven heat transfer components are now analyzed in more detail.

D.2.1 Net Radiation (*Ea*)

Ea represents the sum of short-wave radiation absorbed and reflected by the water surface, long-wave radiation absorbed and reflected by the water surface, and long-wave radiation emitted from the water surface during a given period. Many researchers give detailed methods for calculating net radiation (Beard and Johns, 1985; Raphael, 1962).

Although net radiation is a common energy component, its calculation is not easy. Several tricky factors control radiation, such as air-mass turbidity, shading, and the effect of clouds, which are spatially and temporally highly variable, and these all make the direct calculation difficult, especially for continuous energy balance simulation.

An effective method is to measure radiation in the field. For our study in the Consolidated Drain, a pyranometer has been installed in the field since August 1999 to record the total incoming short-wave radiation, *Easd*. However, incoming and reflected long-wave, and reflected short-wave radiation were not measured (Kieser & Associates, 2002b).

Because of these uncertainties, in this research we have neglected the long-wave radiation loss. In fact, during the day, long-wave radiation may be just a minor portion in the total radiation (Ponce, 1989).

For the reflected short-wave radiation, *Easr* can be calculated as,

$$Easr = al \cdot Easd \qquad (3)$$

where: al = albedo (for wet period, 0.1 (Ponce, 1989); for dry period, 0.15 (Oke, 1978)).

D.2.2 Rain Energy (*Er*)

$$Er = i(t) \cdot A(t) \cdot \Delta t \cdot \rho \cdot c \cdot Tr(t) \qquad (4)$$

where: $i(t)$ = rain intensity (m/s);
$A(t)$ = area of waterbody surface (m^2);
Δt = time interval (s);
ρ = density of water (kg/m^3);
c = specific heat of water $(J / kg \cdot °C)$;
$Tr(t)$ = rainfall temperature $(°C)$.

Rainfall temperature (*Tr*) is another one of the uncertainties. Van Buren (1999) gave a statistical relationship between average air temperature and rainfall temperature, viz. rainfall temperature is 80% of present air temperature. Norman and Roa (2000) found that air temperature (*Ta*) and rainfall temperature could be approximately equal for rain events prior to noon.

In the present research, we use:

$$Tr = (0.8 \sim 1.0) Ta \qquad (5)$$

Thus, rainfall temperature is treated as a candidate parameter for calibration.

D.2.3 Evaporation (*Ee*)

Meyer's equation is a popular method for computing evaporation rate, which needs to be determined prior to the calculation of heat loss by evaporation. The evaporation rate is dependent on several important parameters including Le, e_a, e_o, Hr, and he.

- Latent heat of vaporization, Le (J/kg)

$$Le = 4188(597.3 - 0.57 Tw) \text{ (Bedient and Huber, 1992)} \qquad (6)$$

where: Tw = water temperature $(°C)$

- Saturation vapor pressure at air temperature, e_o (kPa)

$$e_o = 0.611e^{\frac{17.27T_a}{237.3+T_a}} \text{ (Mays, 2001)} \qquad (7)$$

where: Ta = air temperature (°C)

- Actual vapor pressure at air temperature, e_a (kPa)

$$e_a = Hr \cdot e_o \qquad (8)$$

where: Hr = relative humidity (%)

- Evaporation rate, he (m/s)

The commonly used empirical equation, Meyer equation, for evaporation rate is (Ponce, 1989):

$$he = 0.33 \times 10^{-7} con(e_o - e_a)(1 + 0.0447w) \qquad (9)$$

where: w = wind speed at 25ft height (m/s);
con = a constant (15 for small ponds, 11 for large ponds)

Therefore, for small waterbodies,

$$he = 4.96 \times 10^{-7}(e_o - e_a)(1 + 0.0447w) \qquad (10)$$

Practitioners should be careful when using the Meyer equation, because it was derived as a monthly or daily average estimate, and based on ponds or lakes where the waterbody is relatively large and nearly static. Its empirical coefficients should probably be modified and calibrated for low or shallow flow in urban stormwater systems.

Finally, energy loss by evaporation is expressed as:

$$Ee = he \cdot Le \cdot \Delta t \cdot \rho \cdot A \qquad (11)$$

where: A = water surface area (m^2).

D.2.4 Sensible Heat Flux (*Es*)

Sensible heat is conducted into or from a water body by air whenever the temperatures of the water and overlying air are different. It depends on the temperature difference between air and water as well as wind velocity, and cannot be easily computed or observed (Raphael, 1962). A common computational method uses the Bowen ratio, considering sensible heat to be proportional to evaporation heat:

$$Es = R \cdot Ee \qquad (12)$$

where R is the Bowen ratio (Galli, 1990):

$$R = 6.1 \times 10^{-4} \frac{Tw - Ta}{e_{ws} - e_a} P \qquad (13)$$

where: e_{ws} = saturation vapor pressure at water temperature (kPa);
P = atmospheric pressure (kPa).

In the present research, P is assumed to be constant: 101kPa, close to the standard atmospheric pressure at sea level (ASCE, 1996).

It is possible that Tw is close to Ta, and relative humidity is close to 100%. This situation causes $R \approx \pm\frac{0}{0}$, making it impossible to solve for R. Linsley (1988) addressed monthly Bowen ratios and found that they varied from –0.32 in February to 0.25 in November. We imposed the limits of $-1.0<R<1.0$ in this research for the modeling work.

D.2.5 Inflow and Outflow (*Ein and Eout*)

Inflow and outflow represent flow(s) coming into and out of a hydraulic element, respectively, and will have different forms for different elements. For example, inflow can be base flow, or groundwater, while outflow can include in- or ex-filtration. Infiltration to groundwater and exfiltration always exist in overland flow, channels and ponds. Sewers may also experience in- and ex-filtration.

The expressions for inflow and outflow are similar:

$$E = Q(t) \cdot \Delta t \cdot \rho \cdot C \cdot Tw(t) \quad (14)$$

where: Q(t) = flowrate of flow (m^3/s);
$Tw(t)$ = temperature of the flow at previous time $t\,(°C)$.

Note: at any time, the outflow is assumed to have the same temperature as its source, the control volume.

Generally, the effect of impervious land predominates in urban areas. Therefore, at the present stage of this research, in- and ex-filtration are considered to be negligible and are not included in outflow energy (*Eout*) calculations.

D.2.6 Convection Between Water and Ground (*Eg*)

Between water flow and the ground surface, convection will transfer heat flux. To express it, Newton's Law of Cooling is a convenient tool:

$$q = h \cdot A(Tg - T\infty) \quad (15)$$

where: A = ground area (m^2);
Tg = temperature of ground surface $(°C)$;
$T\infty$ = temperature of overlying flow $(°C)$;

q = convective heat flux between flow and ground (J/s);

h = convection heat-transfer coefficient $(watts/m^2 \cdot °C)$.

Determination of h is a difficult task in many cases. Haq (2001) applied an empirical equation from Schlichting (1951), which was derived from classical boundary layer experiments. However, the effect was unsatisfactory. In the present research dimensional analysis is employed to derive general equations for h, which include a few adjustable coefficients. These coefficients are calibrated against field data. For details, see the dissertation by Li (2003). Table D-1 lists all general equations for h for different hydraulic elements:

Table D-1. Convection heat transfer coefficient summary (citations A-13 to A-24 refer to equations in the dissertation by Li).

Hydraulic elements	Dry periods	Wet periods
Closed conduits	Ignored	$h = F1 \cdot \frac{k}{R} \cdot \Pr^{F3}(\mathrm{Re}_R^{F4} + F2)$ (A-13)
Channels	$h = F5\frac{ku}{\nu}(\Pr^{F7} + F6)$ (A-19)	
Subcatchment		$h = G1 \cdot \frac{k}{R} \cdot \Pr^{F3}(\mathrm{Re}_R^{F4} + F2)$ (A-14) $G1 = e^{(F1+F11\cdot R_h+F14\cdot\lvert Tr-Tg\rvert\cdot i^{F16})}$ (A-15)
Ponds and wetlands		$h = F1'\frac{k}{d}(\Pr^{F3'} + F2')$ (A-24)

Eg involves another heat transfer phenomenon: heat conduction inside the underlying envelope below the waterbody. The underlying envelope is assumed to be a substantial solid body whose surface is in contact with water. Urban surfaces, such as roads and parking lots, contain paving, base, subgrade or subbase, and soil beneath the subgrade. Ponds may have a containing envelope of clay, or muck, and soil below. Inside the envelope, if there are temperature differences or gradients, heat flux will occur. Temperature gradients usually exist. Verspagen (1995) provided a few typical profiles.

As for heat transfer processes inside a water body, if groundwater and infiltration are negligible, the internal heat transfer process can be treated as pure conduction.

The present research employs a lumping method, in which, for a hydraulic element, all physical properties are assumed to be uniform. Accordingly, for conduction inside the underlying body, a one-dimensional conduction analysis is used:

$$\frac{\partial^2 T}{\partial y^2} = \frac{1}{\alpha}\frac{\partial T}{\partial t} \qquad (16)$$

where: T = temperature inside the underlying body (°C);
y = spatial variable (m), here y is depth;
t = time variable(s);
α = thermal diffusivity (m^2/s);

α is a physical parameter based on k, ρ and C:

$$\alpha = \frac{k}{\rho C} \qquad (17)$$

where: k = thermal conductivity of underlying body material (*watts/m °C*);

Contact of the waterbody with its underlying substrates and stratification of the underlying substrates are problematic because of the thermal contact resistance (TCR). TCR results from an imperfect joining of the two different materials and will cause a temperature difference between two layers of different material, for example, pavement base and soil. A key coefficient in determining the TCR is the Contact Coefficient, *hco*, which is extremely difficult to obtain. Only a few empirical values are available for metal materials under limited conditions. In the present research, at the joint, the condition is considered to be an ideally joined body. The mean k and C at the joining surfaces between different material layers are used.

The second-order parabolic partial differential equation can be solved using the finite difference method. Li (2003) gives solution details, including treatment of boundary conditions, in his dissertation.

D.3 Heat Budget Schematics

The four different hydraulic elements defined in this research are described below in simple schematics.

D.3.1 Pond and Wetland

Most ponds and wetlands have a containment layer, which has different physical properties from the underlying natural bed. In fact the containment layer may comprise many layers. The thickness of the underlying layers depends on the vegetation, types of materials (e.g.c muck layer, clogged soil), and the properties of the bed soil.

Figures D-2 and D-3 show the heat budget components for ponds in wet and dry periods, respectively. The advective velocity of the pond water is approximately zero even in wet periods.

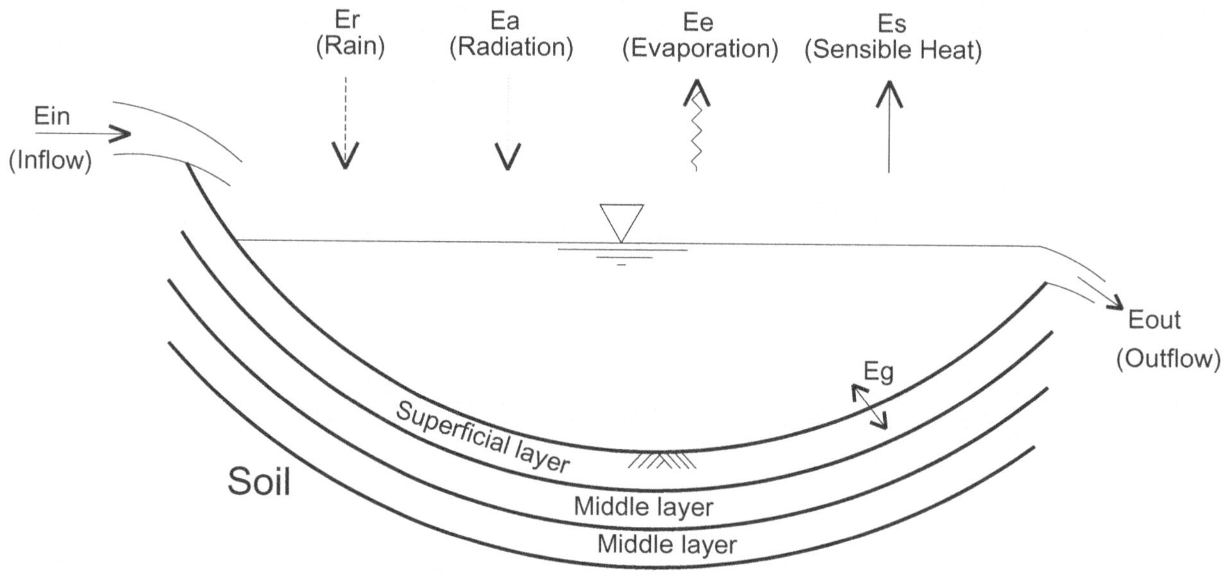

Figure D-2. General sketch of energy budget for ponds during wet periods.

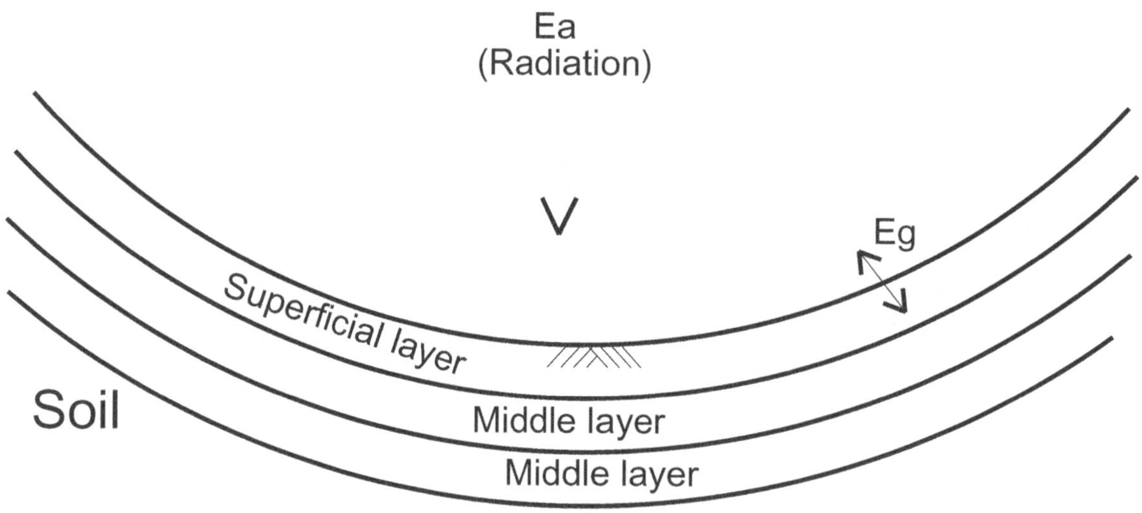

Figure D-3. General sketch of energy budget for dry ponds.

D.3.2 Overland Flow

Pavements commonly constitute a large portion of the urban impervious area. Usually pavements comprise paving and base (subgrade). Figures D-4 and D-5 illustrate the heat budget for paving in wet and dry periods, respectively.

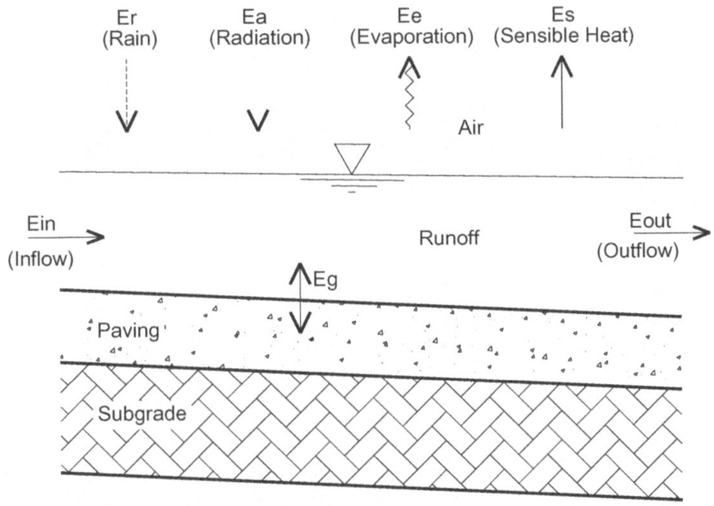

Figure D-4. General sketch of energy budget for wet paving.

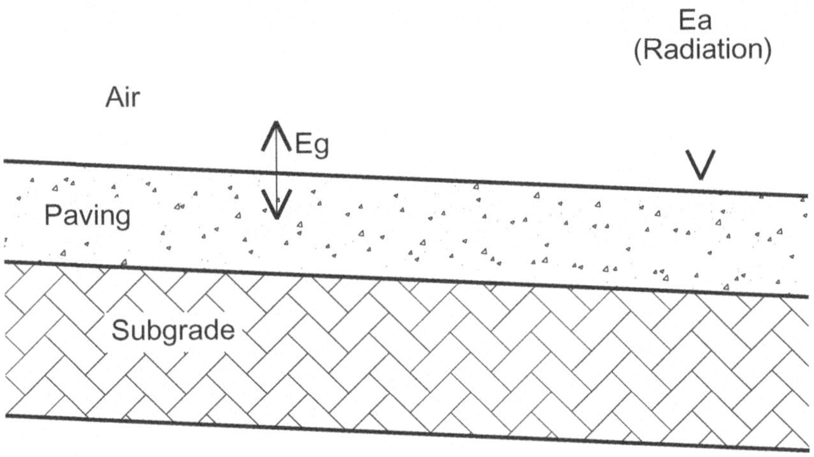

Figure D-5. General sketch of energy budget for dry paving.

D.3.3 Open Channel Flow

Open channel flow has the same energy budget shown in Figures D-4 and D-5, but with different underlying layers.

D.3.4 Closed Conduit Flow

Again, possible biological and chemical reactions in the waterbody are not considered. Further, at this stage of the present research, the circumferential heat transfer inside the pipe wall is neglected. In fact, only the portion of wall in contact with water is considered.

Note that the ambient, underlying body contains pipe wall, and many layers such as bedding and backfill, as illustrated in Figure D-6 (modified, based on McGhee [1991]).

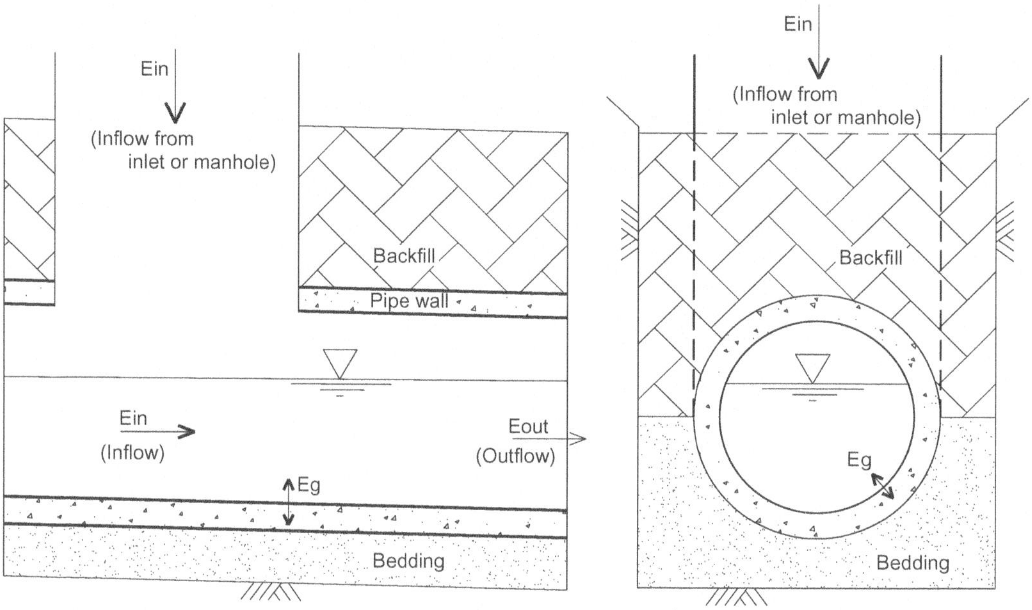

Figure D-6. General sketch of energy budget for partly filled closed conduit.

Supplement: a large number of trials showed that, apart from inflow, outflow and convection, there are extra heat transfer component(s) inside the conduit. The extra component(s) usually decrease water temperature in some way, i.e. a heat loss. A possibility is evaporation, which may occur in a pipe network, as long as sufficient air circulates in the network. The heat loss is assumed to be evaporation in this research.

An expression is created to evaluate the evaporation term:

$$Tnew = [F12 + (1 - F12)\frac{d}{D}] \cdot Told \quad (18)$$

where: $F12$ = an empirical coefficient;

d = water depth (m);

D = pipe diameter (m);

$Told$ = water temperature based on Figure D-6;

Tnew = adjusted water temperature based on equation. 18.

We first calculate the water temperature based on Figure D-6; then use equation 18 to adjust the calculated water temperature.

Equation 18 links water depth to the evaporation term. The more shallow the flow, the more heat losses there will be. In other words, this loss will not occur in a full pipe. F12 is determined by calibration. If this loss is considered negligible, we can simply set F12 = 1.0.

D.4 Heat Transfer Model

Normally, to simulate a stormwater drainage system, the watershed is discretized into many sub-watersheds. The fundamental overland flow elements are treated mathematically as being rectangular in planform, and are termed subcatchments in PCSWMM. Subcatchments are connected via sewers (closed conduit elements), channels (open channel elements), and ponds and wetlands. To compute water temperature in the network, temperature changes in every hydraulic element can be considered separately. A heat transfer model, named HEATRAN, was developed to combine all heat transfer components.

D.4.1 Basic Considerations

Basic requirements and assumptions for the model are:

1. The discharge hydrograph is computed using PCSWWM2002, using the Runoff, Transport and Storage modules;

2. HEATRAN creates runoff storage data in every element, based on inflow and outflow, for the energy balance analysis, using the following relation:

$$S(n+1) = S(n) + [\frac{I(n+1)+I(n)}{2} - \frac{O(n+1)+O(n)}{2}] \cdot \Delta t \quad (19)$$

where: $S(n+1)$ = storage of a given element at time $n+1$;
$S(n)$ = storage at time n;
$I(n)$ = inflow of the element at time n;
$I(n+1)$ = inflow at time $n+1$;
$O(n)$ = outflow of the element at time n;
$O(n+1)$ = outflow at time $n+1$.

3. HEATRAN computes water temperature and temperature profiles of the underlying body in every type of element at every time step;

4. HEATRAN connects all elements, and transfers flow and temperature data at interconnections;

5. Pervious areas are assumed to not contribute direct runoff. Only the impervious areas produce direct runoff;

6. If any water is present, net radiation will only reach the upper water surface, regardless of how shallow the water is. This assumption prohibits penetration of radiation through the water, and radiation will not directly reach the underlying surface, even though in reality the underlying pavement surface can indeed absorb insolation when wet. This assumption renders the heat transfer computational structure clearer and easier to program.

D.4.2 Element Treatment Details

As stated earlier, HEATRAN defines four hydraulic elements: 1) subcatchment; 2) closed conduit; 3) open channel; and 4) pond/wetland.

D.4.2.1 Subcatchment Element

Subcatchments have only downstream elements. In other words, all runoff from a subcatchment flows into channel, closed conduit or pond elements.

Since only impervious areas produce direct runoff, they are a required input for subcatchments. Impervious areas undergo convective heat transfer between fluid flow and paving. Impervious area = total area of subcatchment imperviousness ratio.

For the simulation, outflow comes from PCSWMM (James and James, 2000a&b) output, while rainfall is the inflow. Similar to PCSWMM Runoff, the flow depth is derived from the outflow using Manning's Equation:

$$Q = \frac{Ac}{n} Rh^{\frac{2}{3}} \sqrt{S} \quad (20)$$

where: Q = outflow discharge (m^3/s);
Ac = cross-section area of runoff flow (m^2);
n = roughness;
S = longitudinal slope;
Rh = hydraulic radius (m).

For shallow sheet flow, the hydraulic radius is commonly approximated as d=Rh, where d is the flow depth. Equation 20 can be rearranged to give:

$$d = (\frac{Qn}{B\sqrt{S}})^{\frac{3}{5}} \quad (21)$$

where: B = subcatchment (or runoff flow) width (m);

The depth is also used to obtain flow velocity and the Reynolds number.

D.4.2.2 Channel Flow

Each channel is simplified into a prismatic channel with a uniform longitudinal slope. The cross-section is simplified into a symmetrical trapezoid, with two side slopes. For a given channel, outflow at the downstream end will be the direct input from the PCSWMM Transport module. At the upstream end, all outflow discharges from upstream elements are combined to form the inflow. Both ends contribute to create storage in the channel element.

All water in a single channel element is considered to undergo complete mixing. Convective heat transfer occurs on the water-contact area, the value of which is the product of channel length and wetted perimeter of the channel element. The wetted perimeter is the average of that of the upstream and downstream ends.

Because PCSWMM Transport does not provide flow depth, the depth is back-calculated using channel geometry properties, Manning's roughness and flow. Subsequently, the wetted perimeter and flow velocity are obtained using the depth.

PCSWMM Transport employs kinematic wave theory to compute the discharge hydrograph (James and James, 2000a&b). In kinematic wave theory, at every location for any moment, flow is taken to be uniform, though in reality the flow is unsteady. In order to obtain the flow depth, the uniform flow assumption is again applied. Although the normal depth for uniform flow can be estimated by iteration, for a regular trapezoidal channel the iteration procedure is time-consuming. Therefore, for this study, we have used a tabulation-interpolation method to accelerate the depth estimation (see Li (2003) for details).

Heat conduction through the underlying body depends on the bed materials beneath the channel bottom. Channel elements can be connected to any type and number of elements at their downstream and upstream ends.

D.4.2.3 Closed-Conduit Element

The theoretical treatment of the closed conduit flow element is basically similar to that of the open -hannel element, except for the calculation of normal depth. A tabular relation is used to find the normal depth for a given discharge, as shown in the dissertation by Li (2003).

The underlying body structure depends on the installation, viz. trenching, bedding, and backfilling. Like channel elements, closed conduit elements can be connected to any type and number of elements at the downstream and upstream ends. Dry-period simulation is ignored, as explained in the dissertation by Li (2003).

D.4.2.4 Pond and Wetland Element

A depth-storage-surface relationship is required for pond and wetland simulation. The PCSWMM Storage module uses the relationship to create the treated outflow hydrograph. Because of the static-status assumption, the relationship also becomes the most important factor in pond/wetland simulation.

Convection heat transfer between water and pond/wetland bed will occur on the water-contact surface, which is larger than the surface area of the pond/wetland. In order to obtain this contact area, ponds and wetlands are considered to be conic bodies. The input depth-surface relationship, is fitted to be a standard staged conic body, ignoring a possible inconsistency between the storage and the conic body, as shown in Figure D-7.

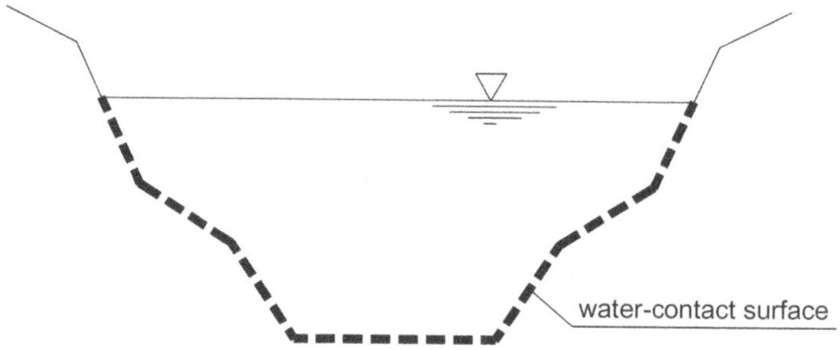

Figure D-7. Fitted pond/wetland cross-section.

The water-contact surface is computed based on geometrical properties of the staged conic body. Using this contact area, an average depth *da* is obtained:

$$da = \frac{Vs}{As} \quad (22)$$

where: As = superficial area of storage volume below water surface (m^2);
Vs = water volume below water surface (m^3).

The underlying surface depends on the hydrographic survey details.

Pond and wetland elements can also be connected to any type and number of elements at the downstream and upstream ends, using dummy connectors if necessary.

D.5 Programming and Operation

All heat transfer computations were programmed into HEATRAN, using the MatLab language. The general operation is:

1. Simulate the watershed hydrology using PCSWMM to obtain a discharge hydrograph for all hydraulic elements—the RUNOFF module is used for subcatchment, TRANSPORT for closed conduit and channel, and STORAGE for pond elements.

2. Transfer the hydrographs computed by PCSWMM to HEATRAN. Since an automatic data transfer is still unavailable, this step is at this time completed manually, using Microsoft Excel.

3. Run HEATRAN, and calibrate candidate parameters to obtain a reasonable result.

D.5.1 Model Structures

HEATRAN comprises six parts:

- Part 1 is a program group, composed of thirteen independent subroutines, which provide the basic physical properties of water and air, and a relationship between normal depth and discharge for a circular channel.
- Part 2 describes the physical characteristics of hydraulic elements, as well as underlying bodies. Simulation parameters are input here, such as time and spatial steps, simulation duration, accuracy control. It also connects them into a network.

- Part 3 collects meteorological data from field gauges, various initial values, and hydrograph data from PCSWMM. If the time step of the gauge is different from the step required by the user, the data will be rearranged to be consistent with the desired time step.
- Part 4 initializes all available information to prepare the final heat transfer computation. Here the flow depth and velocity are back-calculated with discharge values for various elements. A storage file is automatically created for every hydraulic element.
- Part 5 computes all heat transfer processes, as the core unit in the whole program. It discriminates all elements according to the network structure, and picks up all upstream inflows for an element, thus guaranteeing correct heat sources.
- Part 6 provides output options: graphs can be plotted, and data can be output at various time steps.

D.5.2 Calculation Procedure for Heat Transfer

Direct applications of the heat transfer equations may encounter problems, especially for overland sheet runoff flow, which is extremely sensitive to external conditions, due to its small depth.

In the present study, radiation data were collected by a pyranometer installed in a wetland. However, it recorded only the total incoming short-wave radiation. No long-wave radiation or reflected short-wave radiation is available. Due to the computational difficulty, net radiation is just approximated in HEATRAN.

Furthermore, we face another unsolved problem: penetration of the short-wave radiation in a water body. Depending on the color and clarity of the water, some of the incident radiation will be felt at depth as far as 10-15m below the surface (Fischer et al. 1979). This implies that short-wave radiation can penetrate the shallow water depth of overland flow in an urban stormwater system, and reach the ground surface. As a result, some solar heat flux will be absorbed by the ground surface. Fischer et al. (1979) gave a general relationship for diffusion of the solar radiation:

$$\frac{Ey}{Easd} = e^{-\eta(H-y)} \quad (23)$$

where: Ey = solar radiation remaining at a depth y;
$Easd$= total incoming short-wave radiation;
η= an extinction coefficient;
H= water depth.

The relationship is available for some lake studies. However, studies on shallow flows are lacking.

Additionally, equations for evaporation and sensible heat are highly empirical. Particularly, the Meyer equation (9) was derived from experiments on ponds and lakes where the water body is nearly static. Its empirical coefficients will probably not be applicable to low or shallow flow in urban stormwater systems.

In summary, many uncertainties exist, adding to the difficulties of combining all heat transfer components. For example, the above three problems (no long-wave radiation or reflected short-wave radiation data available; uncertainty on the penetration of the short-wave radiation in a water body, and; the highly empirical nature of coefficients used in the Meyer equation) may lead to propagated errors, i.e. erroneously estimated values of input parameters (James, 2002). Considering these uncertainties, and also based on a number of numerical tests, we have imposed several important restrictions in heat transfer computations:

1. Water and ground surface temperatures may not vary abruptly within a short time. In other words, within a short time, temperature will not experience an extreme change; for example, runoff water temperature may not vary from 20°C to 40°C in just 30 s.

2. Water and ground surface temperatures may not reach extremes. For example, experience tells us that, in summer time, a runoff temperature of 50°C or higher is impossible. Similarly, 5°C or lower is also unlikely on a normal summer day.

3. Evaporation (plus sensible heat) or convection may not cause an extreme change on water temperature within a time step. Consider a runoff flow on an insulated surface, with radiation assumed to be zero. In this situation, the only heat loss is evaporation (plus sensible heat). If the air temperature is around 22°C, this heat loss cannot cool the water down to 5°C on a usual summer day.

To provide for these constraints, two approaches are used. First, an allowable range is arbitrarily imposed on the temperature calculation, called *alvar*, which requires that, for an element, after a time step, the new temperature must not exceed a limit derived from the old temperature at the previous step:

$$\left|Tnew - Told\right| \leq al\,\mathrm{var} \cdot Told \qquad (24)$$

This is an arbitrary limit, based on perception and tests, for example, 0.04. The adjustable parameter can initially control an undue change.

Second, a special step-calculation method is used to separately control different heat transfer terms. It includes four steps, as shown in Figure D-8:

- **Step 1.** A simple heat transfer computation is carried out in every element at each time step. This computation only considers rainfall and runoff, assuming no radiation, evaporation (plus sensible heat) or convection. Obviously, the computed water body temperature, *Tw1*, responds only to the variation of rainfall temperature and its own volume.
- **Step 2.** Radiation is added to the temperature calculation. A calibration factor, *F8*, is given to net radiation. As a result, the applied net radiation becomes: $F8 \cdot Ea$. This step produces a new temperature, *Tw2*, usually larger than *Tw1*. If *Tw2* appears abnormal, i.e. exceeding the allowable varying range from *Tw1*, *F8* will be automatically reduced by a small fraction, *mamo,* which is determined by the user, for example, *mamo* = 0.1. This modification is repeated until no abnormal temperature change appears.
- **Step 3.** Evaporation (plus sensible heat) is introduced. Similarly, it also has a calibration factor, *F9*. A new temperature *Tw3* is produced. If *Tw3* is abnormal, *F9* will also be reduced

by *mamo*. Thus the influence of evaporation (plus sensible heat) is attenuated. This modification is repeated until no abnormal temperature change appears.

- **Step 4.** Convection is finally added to the calculation, producing another temperature *Tw4*. The convection formula is an original development in this study, having several calibration coefficients. *F1* will be directly modified as in the previous steps, if necessary. *Tw4,* which is designated to be the final water temperature of this element, will be used for the next step. (Note: for a pipe element, if evaporation is considered, *Tw4* will be treated according to equation (18)).

The entire calculation is time-consuming for long simulations or multi-element watersheds, and manual manipulation may be needed to modify some unreasonable parameters. However, the calculation result is continuous and stable, without any abrupt changes, and the output appears to be reasonable. Figure D-8 also illustrates the complete calculation process for a given element from time step *n* to *n+1*. (*Tw1* and *Tw2* should be ignored for closed conduit elements.) Table D-2 gives the main parameters necessary for every step.

Note that the factors, *F8*, *F9*, as well as *alvar*, are absolutely arbitrary. They are introduced to obtain more reasonable results for low flows. These factors can only be calibrated against field data. If the data are absent, adjustment will depend on user's experience.

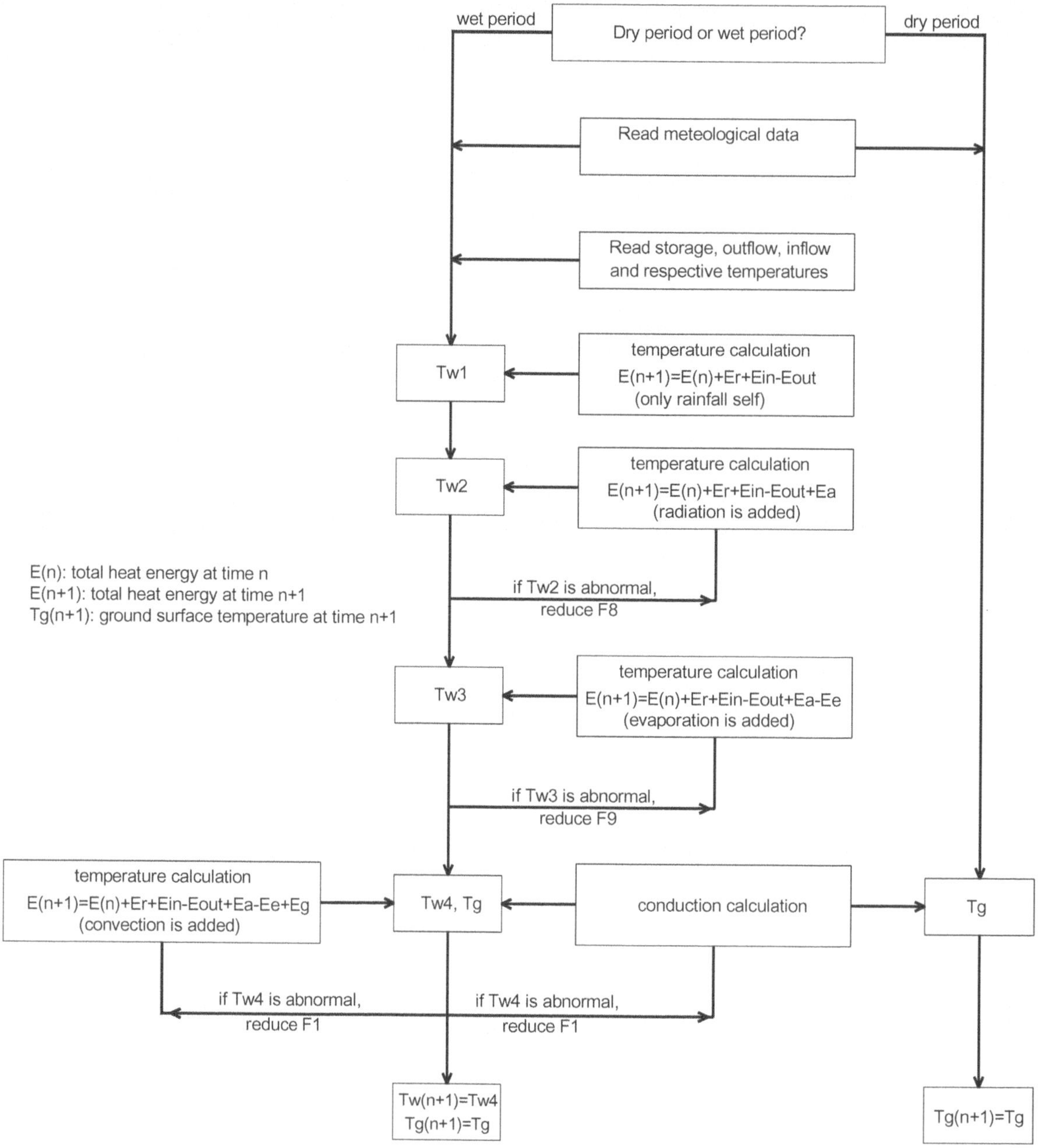

Figure D-8. Step-calculation procedure.

Table D-2. Main parameters needed for step calculation.

Temperature step	Main parameters needed
Tw1 (considering: rainfall)	Rainfall intensity
	Physical size of hydraulic element
	Flow (in and out)
Tw2 (considering: rainfall + radiation)	Radiation
	Coefficient *F8*
Tw3 (considering: rainfall + radiation + evaporation)	Humidity
	Air temperature
	Wind velocity
	Coefficient *F9*
Tw4, Tg (considering: rainfall + radiation + evaporation + convection)	Temperature profile inside underlying body
	F1,F2,F3,F4,F5,F6,F7,F11,F14,F16

REFERENCES

American Iron and Steel Institute. 1996. *Modern Sewer Design, Third Edition.* American Iron and Steel Institute.

American Society of Civil Engineers (ASCE). 1999. *GIS Modules and Distributed Models of the Watershed.* ASCE Task Committee on GIS Modules.

American Society of Civil Engineers (ASCE). 1996. *Hydrology Handbook.* American Society of Civil Engineers.

Barber, E.S. 1957. *Calculation of Maximum Pavement Temperature from Weather Reports.* Bulletin 168. Highway Research Board, Washington, D.C. pp: 1-8.

Beard, J.B. and Johns, D. 1985. The Comparative Heat Dissipation from Three Typical Urban Surfaces: Asphalt, Concrete, and a Bermuda Grass Turf. In *Texas Turfgrass Research.* Texas Agricultural Experiment Station. PR-4329. College Station, TX. pp. 125-133.

Bedient, P.B., and Huber, W.C. 1992. *Hydrology and Floodplain Analysis.* Addison-Wesley Publishing Company. 692 pp.

Brown, R.B. and Otten, L. 1988. A Method to Determine Thermal Properties of Grain from System Response Analysis of a Packed Bed. *International Journal of Thermophysics.* Vol. 9, No. 6.

Chow, V.T. 1959. *Open-Channel Hydraulics.* McGraw-Hill Book Company, Inc., pp:625-627.

Chow, V.T. 1964. *Handbook of Applied Hydrology.* McGraw-Hill Book Company, Inc., pp:11-12.

Dake, J.M.K. 1972. *Essentials of Engineering Hydraulics.* Macmillan. p:91.

Fischer, H.B., List, E.J., Koh, R.C.Y., Imberger, J., and Brooks, N.H. 1979. *Mixing in Inland and Coastal Waters.* Academic Press, Inc. London Ltd. pp: 164-165.

Galli, J. 1990. *Thermal Impacts Associated with Urbanization and Stormwater Management Best Management Practices.* Metropolitan Washington Council of Governments, Washington, D.C. pp. 157.

Haq, R.U. 2001. Thermal Enhancement of Stream Temperature by Urban Stormwater in Portage Creek, MI. MEng Thesis. University of Guelph, Guelph, Canada. 103 pp.

Holman, J.P. 1986. *Heat Transfer.* McGraw Hill, Toronto. 676 pages.

Heathcote, I.W. 1998. *Integrated Watershed Management.* John Wiley & Sons, Inc. pp: 41.

Isaacson, E. de St Q. and Isaacson, M. de St Q. 1975. *Dimensional Methods in Engineering and Physics: Reference Sets and the Possibilities of their Extension.* John Wiley & Sons, Inc. pp: 3,91.

Jain, S.C. 2001. *Open-Channel Flow.* John Wiley & Sons, Inc. 328 pp.

James, W. 2002. *Rules for Responsible Modeling, 3rd edition.* Computational Hydraulics International. pp: 152.

James, W. and James, R. 2000a. *Water Systems Models Hydraulics: A guide to the Extran, Transport and Storage Treatment modules of the USEPA SWMM4.* Computational Hydraulics International, Guelph, Canada.

James, W. and James, R. 2000b. *Water Systems Models Hydrology:A guide to the Rain, Temperature and Runoff modules of the USEPA SWMM4.* Computational Hydraulics International, Guelph, Canada.

Kieser & Associates (K&A) 2002a. Information Sheets (by E-mail, express mail or website: http://www.kieser-associates.com)

Kieser & Associates (K&A) 2002b. *Progress Report #4, Stormwater Thermal Enrichment in Urban Watersheds.* Kalamazoo, MI.

Kunz, K.S. 1957. *Numerical Analysis.* McGraw-Hill Book Company, Inc.

Lau, Y.L. 1971. *Temperature Distribution Due to the Release of Heated Effluents into Channel Flow.* Technical Bulletin No. 55, Inland Waters Branch, Department of the Environment, Ottawa, Canada.

Li, Yi. 2003. Thermal Enrichment by Urban Stormwater of Receiving Water Systems. Master's thesis. School of Engineering, University of Guelph, Ontario, Canada. May 2003.

Linsley, R.K. 1988. *Hydrology for Engineers.* McGraw-Hill, Inc. pp: 131.

Mays, L.W. 2001. *Water Resources Engineering.* John Wiley & Sons, Inc. 761 pages.

McGhee, T.J. 1991. *Water Supply and Sewerage.* McGraw-Hill, Inc. pp: 306.

National Academy of Sciences and National Academy of Engineering. 1972. *Water Quality Criteria.* Washington, D.C. pp: 594.

Norman J.M. and Roa, A. 2000. *Modeling the Effects of the Natural Environmental and Urban Runoff on Stream Temperature.* Final Report. June 22, 2000.

Oke, T.R. 1978. *Boundary layer climates.* Methuen & Co Ltd. pp: 247.

Picksley, J.W., and Deletic, A. 1999. The Thermal Enrichment of Storm Runoff from Paved Areas – a Statistical Analysis. In *New Applications in Modeling Urban Water Systems*. Monograph 7 in the series. W. James, Ed. Computational Hydraulics International, Guelph, Canada.

Pluhowski, E.J. 1970. *Urbanization and Its Effect on the Temperature of the Streams on Long Island, New York.* U.S. Geological Survey Professional Paper 627-D. pp: D1-D59.

Ponce, V.M. 1989. *Engineering Hydrology: Principles and Practices*. Prentice-Hall, Inc.

Raphael, J.M. 1962. Prediction of Temperature in Rivers and Reservoirs. *Journal of the Power Division*. Proceedings of the American Society of Civil Engineers. Vol: 88, pp: 157-181.

Roa-Espinosa, A., Norman J.M., Wilson, T.B. and Johnson, K. *Predicting the Impact of Urban Development on Stream Temperature Using TURM (Temperature Urban Runoff Model).* (obtained from one of the authors by e-mail).

Rosenberg, D.U.Von. 1969. *Methods for the Numerical Solution of Partial Differential Equations*. American Elsevier Publishing Company, Inc. pp: 19.

Schlichting, H. 1951. *Boundary-Layer Theory*. McGraw-Hill, Inc, 1979. pp: 276.

Schroeter, H.O. Van, Vliet, D.J., Boehmer, K., and Beach, D. 1995. Continuous In-stream Temperature Modeling: Integration with a Physically based Subwatershed Hydrology Model. In *Advances in Modeling the Management of Stormwater Impacts.* James, W. Ed. Computational Hydraulics International, Guelph, Canada.

Thomann, R.V., and Mueller, J.A. 1987. *Principles of Surface Water Quality Modeling and Control.* Harper & Row, Publishers, Inc. pp: 599-600.

Van Buren, M.A.V. 1999. Thermal Enhancement of Urban Receiving Waters. PhD thesis, Queen's University at Kingston, Canada. 199 pages.

Velz, C.J., and Gannon, J.J. 1960. Forecasting Heat Loss in Ponds and Streams. *Water Pollution Control Federation*. Vol: 32, pp: 392-417.

Verspagen, B. 1995. Experimental Investigation of Thermal Enrichment of Storm Water Runoff From Two Paving Surfaces. MSc Thesis, University of Guelph, Guelph, Canada. 163 pages.

Weatherbe, D.G. 1995. A Simplified Stream Temperature Model for Evaluating Urban Drainage Inputs. In *Modern Methods for Modeling the Management of Stormwater Impacts*. James, W. Ed. Computational Hydraulics International, Guelph, Canada. pp. 259-273.

Xie, D.M. 1993. Modelling Solar Thermal Enrichment of Streams Due to Urban Storm Water. MSc Thesis, University of Guelph, Guelph, Canada. 118 pages.

http://www.city.fitchburg.wi.us/Stormwater/CH27.pdf Erosion Control and Stormwater Management Ordinance (adopted August 13, 2002), City of Fitchburg, WI.

http://www.co.dane.wi.us/commissions/lakes/doc/summaryaug2002.doc Erosion Control and Stormwater Management Ordinance (August 22, 2002), Dane County, WI.

http://www.co.dane.wi.us/landconservation/homebnd.htm Land Conservation Department, Dane County, WI.

http://www.westchestergov.com/planning/environmental/FAQsheets/FAQ5.pdf County Committee on Nonpoint Source Pollution, Westchester County, NY

http://www.wiik-hoeglund.com/products/hdpe.html Typical properties of HDPE, Wiik & Hoeglund Public Company Limited, Thailand

APPENDIX E

2003 DATA ANALYSIS

Funding for this research provided by the Water Environment Research Foundation supported full field efforts through the 2002 monitoring season (early October 2002). Kieser & Associates continued the operation of some key monitoring instruments in the study area after 2002. Thermistors 3, 4, and 5 and rain gauge 8 (see Chapter 3.0 and Appendix B) continue to provide important data for further understanding the thermal behavior of the Consolidated Drain stormwater treatment system. After the completion of the primary data analysis for this report, year 2003 data from these select instruments are provided here to augment previous data. This appendix provides a preliminary analysis of the new data.

E.1 Overview

As concluded in Chapters 5.0 and 6.0, vegetative cover plays an important role in mitigating stormwater thermal enrichment in the treatment system. It is also crucial in maintaining the thermal regime in the two drain channels and the receiving stream. Summer of 2003 marked the third growth season since the completion and operation of the Consolidated Drain Stormwater Quality and Trailways Facility. As the planted and natural vegetation matured in the system, particularly in the Consolidated Drain and Schuring Drain channels, it was expected that thermal impact of runoff from the Consolidated Drain watershed on Portage Creek would diminish in the summer months of 2003 compared to previous years.

E.2 Data Analysis

Data from 2003 for thermistors 3, 4, and 5, and rain gauge 8 are analyzed here from two perspectives: thermal regime change in Portage Creek and diel fluctuations in the Drains and the Creek. To be comparable to the datasets from previous years, the August-September two month comparative time period is again targeted here for data interpretation. Due to a mid-September 2003 thermistor malfunction in the Consolidated Drain (T3), late September data are limited in this analysis.

Because of the “reference” function of the T4 data in the analysis of thermal regime change in Portage Creek (see Section 6.4.1), the loss of T4 to vandalism in 2002 made it difficult to make meaningful interpretation of the T3 and T5 data in terms of the effect of stormwater runoff from the Consolidated Drain watershed on Portage Creek. To understand how the maturing vegetation in the treatment system drains affects the thermal treatment efficiency of the system, 2003 data are compared to 2001 data.

The main difference between 2001 and the later post-construction years (2002 and 2003), besides vegetation growth and other obvious physical and biological changes (e.g., accumulation of sediment in the treatment cells), is the additional flow from the fountain of the reflecting pond entering the Consolidated Drain channel (Figure 3-2). As noted in Chapter 6.0 (Section 6.4.2), this flow was at about 0.028 m^3/s and existed only during daylight hours, as the fountain did not operate during the night. Because of the exposure of this flow component to direct solar radiation in the drain channel (see Section 4.3.6), it became a major source of diel temperature fluctuation in the Consolidated Drain in 2002 during dry weather conditions. In 2003, the portion of the Consolidated Drain channel where this fountain flow is discharged saw more vegetative growth both in and around it. As a result, any significant change in the thermal regime in downstream flow components compared to 2002 (or 2001) would be a clear indication of the thermal benefit of vegetative cover.

E.2.1 Thermal Regime Change

Temperature differences between Portage Creek upstream of the Consolidated Drain confluence (T4) and the Consolidated Drain (T3), and Portage Creek downstream of the confluence (T5) are illustrated in Figure E-1 for 2003. This plot is comparable to Figures 6-2 (2000 data) and 6-3 (2001 data). It shows the impact of the Consolidated Drain flow on the

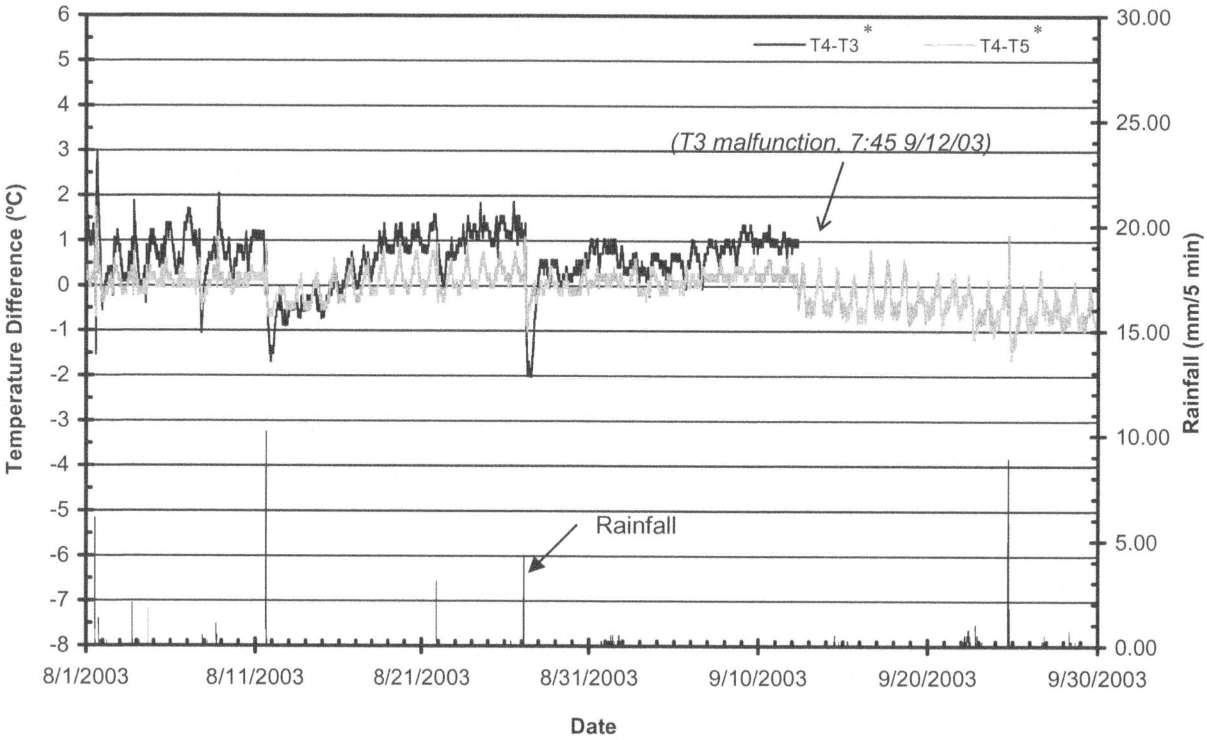

Figure E-1. Water temperature differences between the Consolidated Drain channel and Portage Creek in the months of August and September in 2003 (two years after the stormwater treatment system came on-line). * Temperature differences between thermistors T4 and T3, and between thermistors T4 and T5.

thermal regime of Portage Creek. In general, the stormwater treatment system was able to mitigate the acute thermal impact from storm runoff (fewer and smaller down spikes noted for T4-T3). Statistical analysis of the temperature differences between Portage Creek upstream and downstream of the Consolidated Drain discharge (T4-T5) for all the fours years (with complete record for August through September) are provided in Figure E-2 and Table E-1. These statistical descriptions show clear signs of milder thermal impact of the Consolidated Drain to Portage Creek. First, there were fewer extreme downstream temperature increases in 2003 than

2001, indicated by the more closely packed box plot for 2003 and the smaller data range value (Table E-1). Second, the median value of the 2003 dataset was higher than that of the 2001

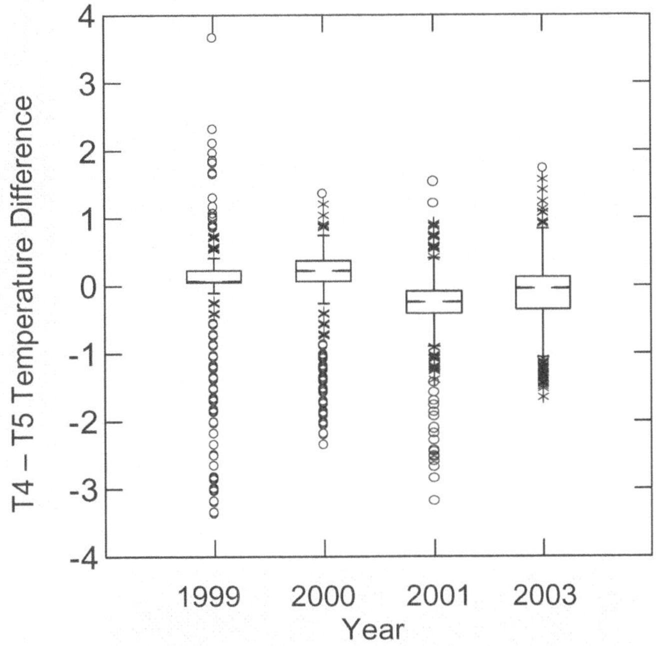

Figure E-2. Box plots of the water temperature difference in Portage Creek upstream (T4) and downstream (T5) of the confluence with the Consolidated Drain for the months of August and September from 1999-2003 (comparable 2002 data not available).

Table E-1. Basic statistics of the 5-minute water temperature difference (ΔT=T4-T5) in Portage Creek upstream (T4) and downstream (T5) of the confluence with the Consolidated Drain channel for the months of August and September in 1999, 2000, 2001, and 2003.

	1999	2000	2001	2003
N of cases	17,272	17,281	17,281	17,568
Minimum (°C)	-3.361	-2.350	-3.180	-1.650
Maximum (°C)	3.667	1.361	1.530	1.730
Range (°C)	7.028	3.711	4.710	3.380
Median (°C)	0.067	0.222	-0.240	-0.040
Mean (°C)	0.057	0.157	-0.226	-0.073
95% CI upper [a]	0.062	0.163	-0.222	-0.067
95% CI lower [a]	0.051	0.152	-0.230	-0.079
Variance	0.118	0.137	0.088	0.146
Skewness	-4.051	-2.967	-0.568	-0.499
Kurtosis	26.775	11.849	4.705	0.122

[a] 95% confidence interval upper and lower limits.

dataset, indicating a lower overall downstream temperature (T5) in 2003. Third, skewness and kurtosis values indicate a more normally distributed dataset in 2003 than any previous years, further pointing to a more stable Portage Creek thermal regime.

A continued trend of smaller T4-T3 as shown by Figures 6-2 and 6-3 is noted in Figure E-1. From 2000 to 2001 and then to 2003 (2002 data not available for T4), the vegetative cover in the Consolidated Drain channel went through a transition from densely vegetated, to new growth, to established growth, respectively. With the vegetation established in 2003, one would expect the water temperature in the Consolidated Drain to be at least lower than that in 2001 (hence higher T4-T3 values). However, the opposite was true. A possible explanation is the warming of the fountain flow by direct solar radiation. Although not so high as in 2002, the exposure of the flow continued in 2003. Compared to other flow components in the Consolidated Drain channel during dry weather conditions, most of which are groundwater baseflow, this fountain flow had the most exposure to direct solar radiation. Consequently, it was a warming source to the Consolidated Drain (see Section 4.3.6). At about 0.028 m^3/s, this flow was a sizeable addition to the Consolidated Drain during dry weather conditions, compared to Schuring Drain's 0.037 m^3/s (> 95% of low flow) and downstream Consolidated Drain's 0.085 m^3/s (> 95% of low flow). Therefore, it is very likely that the fountain flow contributed to higher T3 temperatures in 2003.

E.2.2 Diel Temperature Fluctuations

Although the warming of the fountain flow continued in 2003, the beneficial effect of established vegetation in the Consolidated Drain channel is clearly demonstrated when diel temperature fluctuations are considered. Compared to Figure 6-9, Figure E-3 shows that diel temperature fluctuations in the downstream Consolidated Drain (T3) were much lower (2-4°C) in 2003 than 2002 while the Portage Creek downstream thermistor T5 behaved similarly in both years. As discussed above, the apparent reason is the increased vegetative cover available to the portion of the Consolidated Drain that carried the fountain flow during daylight hours. Although on average the fountain flow was warmed due to exposure to solar radiation, the increased vegetative cover in 2003 was able to dampen the diel fluctuations with lower daily highs.

A closer examination of the temperature readings of T3 from 2003 and 2002 revealed a clear difference in the Consolidated Drain temperature regime in these two post-construction years (Figures E-4 and E-5). The higher diel fluctuations of T3 in 2002 are evident in Figure E-5. In terms of the timing of daily highs and lows, both 2002 and 2003 continued the pattern of 2001, i.e., T3 peaks or valleys occurred at about the same time as T5 and T4/T6, reflecting the fact that shading and flow conditions in the post-construction Consolidated Drain were not the same as those in pre-construction conditions, even after three seasons of vegetative growth. Due to the addition of flows from Schuring Drain and the fountain, the pre-construction thermal regime of the Consolidated Drain that was defined by a small baseflow (< 0.028 m^3/s) completely covered by vegetation cannot be replicated.

Overall, available 2003 data provided by the remaining monitoring equipment indicate the Consolidated Drain stormwater treatment system and redesigned and reconstructed drain/floodway channels were gaining thermal efficiency as a result of continued vegetative growth in the system. In this appendix, an analysis on the effects of vegetative growth on the thermal treatment efficiency of the treatment cells (the ponds and the wetland) is not provided. This is due to: 1) no monitoring equipment was available to measure temperature in the

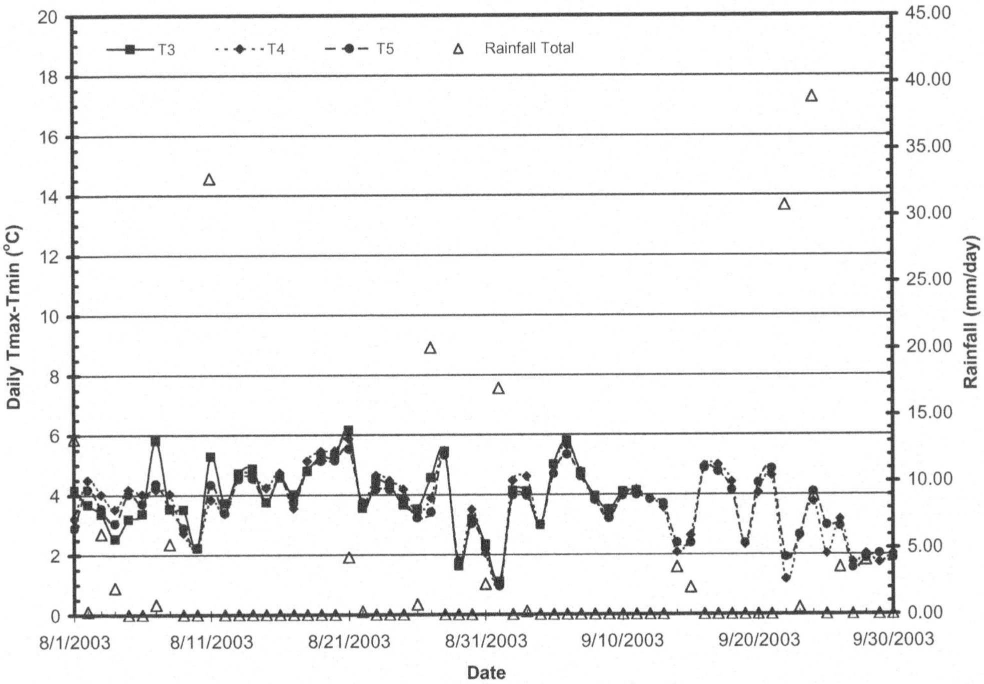

Figure E-3. Diel temperature fluctuations of the Consolidated Drain channel and Portage Creek in the months of August and September in 2003 (two years after the stormwater treatment system came on-line), illustrated by the difference between the daily temperature maximum and minimum. *See Figure 6-2 for thermistors T3, T4 and T5 locations; see also Section 6.4 for more description of thermistor locations.

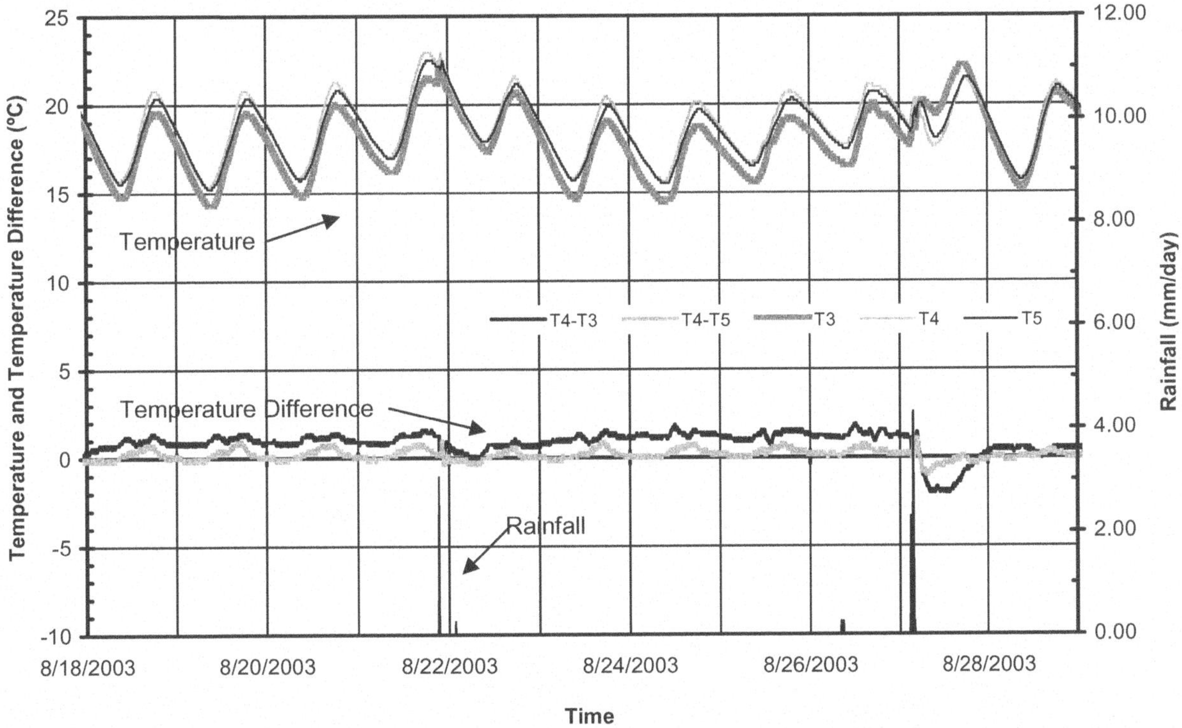

Figure E-4. An example of 5-min interval temperature records of the Consolidated Drain channel and Portage Creek in the months of August and September in 2003 (second full growth season after the stormwater treatment system came on-line), illustrated by the difference between the daily temperature maximum and minimum. *See Figure 6-2 for thermistors T3, T4 and T5 locations; see also Section 6.4 for more description of thermistor locations.

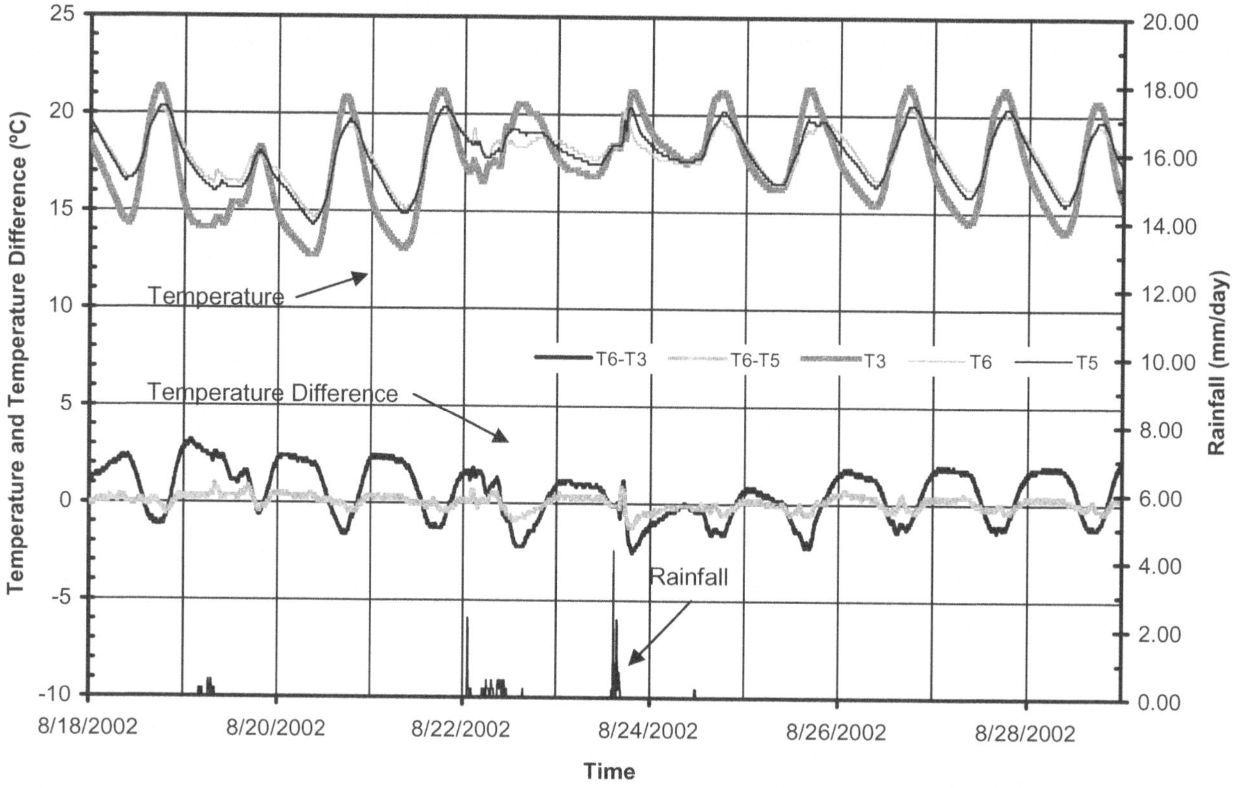

Figure E-5. An example of 5-min interval temperature records of the Consolidated Drain channel and Portage Creek in the months of August and September in 2002 (first full growth season after the stormwater treatment system came on-line), illustrated by the difference between the daily temperature maximum and minimum.

treatment cells; and 2) as discussed in Chapters 5.0 and 6.0, without groundwater input, the most the treatment cells can do is to bring the temperature of treated stormwater to near ambient air/ This was largely achieved by 2002. More vegetative growth may improve the system's treatment efficiency for common pollutant removal but may not likely to help lower the outflow temperature at the end of the treatment cells. Continued canopy growth within and surrounding the channels will likely improve dry-weather temperature moderation of diel fluctuations.

REFERENCES

Bedient, P.B.; Huber, W.C. 1992. *Hydrology and Floodplain Analysis, 2nd ed.* Addison-Wesley Publishing Company.

Booth, D.B.; Jackson, C.R. 1997. Urbanization of Aquatic Systems: Degradation Thresholds, Stormwater Detection, and the Limits of Mitigation. *Journal of the American Water Resources Association.* 33:1077.

Booth, D.B.; Hartley, D.; Jackson, C.R. 2002. Forest Cover, Impervious-Surface Area, and the Mitigation of Stormwater Impacts. *Journal of the American Water Resources Association.* 38:835-845.

Bretz, S.; Akbari, H.; Rosenfeld, A. 1998. Practical Issues for Using Solar-Reflective Materials to Mitigate Urban Heat Islands. *Atmospheric Environment.* 32:95-101.

Chen, Y.D.; Norton, D.J.; Craing, J.P. 1996. Enhancement and Application of HSPF for Stream Temperature Simulation in Upper Grande Ronde Watershed Oregon. *Proceedings of Watershed 1996.*

Claytor, R.A.; Brown, W.E. 1996. *Environmental Indicators to Assess Stormwater Control Programs and Practices.* Final Report. Center for Watershed Protection: Ellicott City, MD.

Collins, W.D. 1925. Temperature of Water Available for Industrial use in the United States. United States Geological Survey Water Supply Paper 520-F.

Crimmins, M.A. 1998. Spatial and Temporal Variations of Cloud-to-ground Lightning in Southern Michigan. M.A. Thesis Western Michigan University, Kalamazoo, MI.

DeNicola, D.M. 1996. Periphyton Responses to Temperature at Different Ecological Levels. In *Algal Ecoloy;* R.J. Stevenson, M.L. Bothwell, and R.L. Lowe, Eds.; Academic Press: New York.

Doll, D.; Ching, J.K.S.; Kaneshiro, J. 1985. Parameterization of Subsurface Heating for Soil and Concrete Using Net Radiation Data. *Boundary-layer Meteorology.* 32:351-372.

Dunne, T.; Leopold, L.B. 1978. *Water in Environmental Planning.* W.H. Freeman and Company, New York.

Eaton, J.G.; Scheller R.M. 1996. Effects of Climate Warming on Fish Thermal Habitat in Streams of the United States. *Limnology and Oceanography.* 41:1109-1115.

Galli, J. 1990. *Thermal Impacts Associated with Urbanization and Stormwater*

Management Best Management Practices. Metropolitan Washington Council of Governments. 157pp.

Goward, S.N. 1981. Thermal Behavior of Urban Landscapes and the Urban Heat Island. *Physical Geography* 2:19-33.

Haq, R.U. 2001. *Thermal Enhancement of Stream Temperature by Urban Stormwater in Portage Creek, MI.* MEng Thesis. University of Guelph. 103pp.

Heaton, S. 1997. *A Biological Survey of Portage Creek in the Vicinity of Plastics Engineering, WL Molding, and Pharmacia & Upjohn, Inc. Kalamazoo County, Michigan.* Report Number MI/DEQ/SWQ-97/046. Michigan Department of Environmental Quality, Lansing, MI.

Heaton, S. 1999. *Loading Assessments of Phosphorus Inputs to Lake Allegan, 1998.* Report Number MI/DEQ/SWQ-99/125. Michigan Department of Environmental Quality, Lansing, MI.

Hinz, L.C.; Wiley, M.J. 1997. Growth and Production of Juvenile Trout in Michigan Streams: Influence of Temperature. Michigan Department of Natural Resources, Fisheries Division Research Report 2041, Ann Arbor, MI.

James, W.; James, R. 2000. *Water Systems Models Hydraulics: A Guide to the Extran, Transport, and Storage Treatment Modules of the U.S. EPA SWMM4*. Computational Hydraulics International. Guelph, Canada.

James, W.; Xie, J.D.M. 1999. Modeling Thermal Enrichment of Streams Due to Solar Heating of Local Urban Stormwater. In *New Applications in Modeling Urban Water Systems*. James, W., Ed.; Computational Hydraulics International, Guelph, Canada. 7:139-157.

James, W.; Verspagen, B. 1997. Thermal Enrichment of Stormwater by Urban Pavement. In *Advances in Modeling the Management of Stormwater Impacts.* James, W. Eds.; Computational Hydraulics International, Guelph, Canada. 5:155-177.

Jensen, M.E.; Burman, R.D.; Allen, R.G. 1990. *Evapotranspiration and Irrigation Water Requirements—ASCE Manuals and Reports on Engineering Practices No. 70.* American Society of Civil Engineers, New York.

Johanson, R.C.; Imhoff, J.C.; Kittle, J.L.; Donigan, A.S. 1984. *Hydrological Simulation Program – Fortran (HSPF): User Manual.* EPA-600/9-80-015. United States Environmental Protection Agency. pp:248-253.

Johnstone, H.C.; Rahel, F.J. 2003. Assessing Temperature Tolerance of Bonneville Cutthroat Trout Based on Constant and Cycling Thermal Regimes. *Transactions of the American Fisheries Society.* 132:92-99.

Kadlec, R.H.; Knight, R.L. 1996. *Treatment Wetlands*. Lewis Publishers, Boca Raton, FL.

Kadlec, R.H.; Reddy, K.R. 2001. Temperature Effects in Treatment Wetlands. *Water Environment Research.* 73:543-557.

Karr, J.R. 1981. Assessment of Biotic Integrity Using Fish Communities. *Fisheries.* 32:21-27.

Kieser, M.S.; Crimmins, M.A.; Jacobson, J. 1999. Thermal Enrichment: A Critical Design Criterion for Protecting and Restoring Cold Water Habitats in North Temperate Climates. Presented at the 19th International Symposium of the North American Lake Management Society, Reno, NV, December 2, 1999.

Kieser & Associates; URS Greiner Woodward Clyde. 1999. *Consolidated Drain No.1 Feasibility Study: An Integrated Approach Focusing on Environmental & Public Benefits* for the City of Portage. Portage, MI. October 27, 1999.

Kieser & Associates. 2002. *WERF Progress Report #4, Stormwater Thermal Enrichment in Urban Watersheds.* Kalamazoo, MI.

Klein, R.O. 1979. Urbanization and Stream Quality Impairment. *Water Resources Bulletin* 15:948-963.

LeBlanc, R.T.; Brown, R.D.; FitzGibbon, J.E. 1997. Modeling the Effects of Land Use Change on the Water Temperature in Unregulated Urban Streams. *Journal of Environmental Management.* 49:445-469.

Li, Y.; James, W. 2003. Model Background Theory. Appendix D of this report.

Magnuson, J.J.; DeStasio, B.T. 1997. Thermal Niche of Fishes and Global Warming. In *Global Warming: Implications for Freshwater and Marine Fish.* C.M. Wood and McDonald, D.G., Eds. Society for Experimental Biology Seminar Series 61. Cambridge University Press, Cambridge, UK.

Matthews, W.J. 1987. Physiochemical Tolerance and Selectivity of Stream Fishes as Related to Their Geographic Ranges and Local Distributions. In *Community and Evolutionary Ecology of North American Stream Fishes.* W. J. Matthews and Heins, D.C., Eds. University of Oklahoma Press, Norman, OK. pp.111-120

Maxted, J.R.; Shaver, E. 1999. The Use of Detention Basins to Mitigate Stormwater Impacts to Aquatic Life. In *National Conference on Retrofit Opportunities for Water Resource Protection in Urban Environments.* EPA/625/R-99/002, U.S. Environmental Protection Agency, Office of Research and Development. pp. 6-15.

Metropolitan Council Environmental Services. 1997. *Lake McCarrons Wetland Treatment System Phase III Study Report.* 99pp.

Meyer, A.F. 1944. *Evaporation from Lakes and Reservoirs.* Minnesota Resources Commission. St. Paul, MN.

Michigan Department of Environmental Quality, Surface Water Quality Division, Great Lakes Environmental Assessment Section. 2002. *Qualitative Biological Habitat Survey Protocols for Wadeable Streams and Rivers. GLEAS Procedure #51.* Lansing, MI. 40pp.

Michigan State University Extension (MSUE). 1999. Water Quality Data Access System. Available at http://www.gis.iwr.mus.edu/storet (last verified September 30, 2003).

National Stormwater Best Management Practices Database. 2002. Available at http://www.bmpdatabase.org (last verified December 2002).

Paul, M.J.; Meyer, J.L. 2001. Streams in the Urban Landscape. *Annual Review of Ecology and Systematics.* 32:333-365.

Penmen, H.L. 1948. Natural Evaporation from Open Water, Bare Soil, and Grass. *Proceedings of the Royal Society of London, Serial A.* 27:779-787.

Picksley, J.W.; Deletic, A. 1999. The Thermal Enrichment of Storm Runoff from Paved Areas. In *New Applications in Modeling Urban Water Systems.* James, W., Ed. Computational Hydraulics International, Guelph, Canada. 7:127-138.

Pluhowski, E.J. 1970. *Urbanization and Its Effect on the Temperature of the Streams on Long Island, New York.* Professional Paper 627-D. United States Geological Survey.

Program Managers: EOA, Inc., Santa Clara Valley Urban Runoff Pollution Prevention Program. 2001. Environmental Indicators: A Useful Tool for Assessing Stormwater Programs? Final Report 96-IRM-3. Water Environment Research Foundation, Alexandria, VA.

Roa-Espinosa, A.; Norman, J.; Wilson, T.; Johnson, K. 2002. *Thermal Impact Analysis of Token Creek Subwatershed and Validation of the Temperature Urban Runoff Model (TURM).* Dane County Land Conservation Department, Dane County, WI.

Roesner, L.A.; Bledsoe, B.P.; Brasher, R.W. 2003. Are Best-Management-Practice Criteria Really Environmentally Friendly? *Journal of Water Resources Planning and Management.* 127:150-154.

Schrank, A.J.; Rahel, F.J.; Johnstone, H.C. 2003. Evaluating Laboratory-Derived Thermal Criteria in the Field: An Example Involving Bonneville Cutthroat Trout. *Transactions of the American Fisheries Society.* 132:100-109.

Schueler, T.R. 1987. *Controlling Urban Runoff: A Practical Manual for Planning and Designing Urban BMPs.* Prepared for the Washington Metropolitan Water Resources Planning Board. Metropolitan Council of Governments, Washington, DC.

Schueler, T.R.; Galli, J. 1995. The Environmental Impact of Stormwater Ponds. *In Stormwater Runoff and Receiving Systems: Impact, Monitoring, and Assessment.* Herricks, E., Ed. CRC Lewis Publishers, Boca Raton, FL.

Strecker, E.W.; Quigley, M.M.; Urbonas, B.R.; Jones, J.J.; Clary, J.K. 2001. Determining Urban Storm Water BMP Effectiveness. *Journal of Water Resources Planning and Management* 127:144-149.

Strecker, E.W.; Quigley, M.M.; Urbonas, B.R. 2002. *How Well Do BMPs Work? A Guidance Manual for Meeting the National Stormwater BMP Database Requirements and an Update on BMP Effectiveness.* National TMDL Science and Policy 2002 Specialty Conference.

Swanson, S.M.; Culp, J.M.; Eipert, S.; Franklin, C.G.J.; Lamberti G.A.; Lowell R.B.; Taylor, B.R.; Van Den Assem, D.J.; da Costa Vargas, S.V. 2001. *Multiple Stressors: Risk-Based Framework and Experimental Design for Cause/Effect Relationships.* Report 00ECO2. Water Environmental Research Foundation, Alexandria, VA.

Taha, H. 1997. Urban Climates and Heat Islands: Albedo, Evapotranspiration, and Anthropogenic Heat. *Energy and Buildings* 25:99-103.

Theurer, F.D.; Voos, K.A.; Miller, W.J. 1984. *Instream Water Temperature Model.* U. S. Fish and Wildlife Service. pp:II.5-II.86.

Thomann, R.V.; Mueller, J.A. 1987. *Principles of Surface Water Quality Modeling and Control.* Harper Collins Publishers Inc.

U.S. Environmental Protection Agency. 1999. *Rapid Bioassessment Protocols for Use in Streams and Wadeable Rivers in Periphyton, Benthic Macroinvertebrates, and Fish, Second Edition.* EPA 841-B-99-002. Washington, DC.

U.S. Environmental Protection Agency. 2001. Better Assessment Science Integrating Point and Nonpoint Sources (BASINS Version 3.0). Available at http://www.epa.gov/OST/BASINS/

U.S. Environmental Protection Agency. 2003. *National Measurement Measures to Control Nonpoint Source Pollution from Urban Areas.* Draft Report. Available at http://www.epa.gov/owow/nps/urbanmm/index.html (last verified November 2003).

U.S. Environmental Protection Agency. http://oaspub.epa.gov/waters/national_rept.control (accessed June 2, 2004).

Van Buren, M.A. 1999. Thermal enhancement of urban receiving waters. Ph.D. thesis, Queen's University, Canada.

Van Buren, M.A.; Watt, W.E., Marsalek, J.; Anderson, B.C. 2000a. Thermal Balance of

On-Stream Storm-Water Management Pond. *Journal of Environmental Engineering.*126:509-517.

Van Buren, M.A.; Watt, W.E.; Marsalek, J.; Anderson, B.C. 2000b. Thermal Enhancement of Stormwater Runoff by Paved Surfaces. *Water Research.* 34:1359-1371.

Vannote, R.L.; Minshall, G.W.; Cummins, K.W.; Sedall, J.R.; Cushing, C.E. 1980. The River Continuum Concept. *Canadian Journal of Fisheries and Aquatic Sciences.* 37:130-137.

Vannote, R.L.; Sweeney, B.W. 1980. Geographic Analysis of Thermal Equilibria: A Conceptual Model for Evaluating the Effect of Natural and Modified Thermal Regimes on Aquatic Insect Communities. *The American Naturalist* 115:667-695.

Verspagen, B. 1995. Experimental Investigation of Thermal Enrichment of Storm Water Runoff From Two Paving Surfaces. MS Thesis, University of Guelph, Canada.

Ward, J.V.; Stanford, J.A. 1979. *The Ecology of Regulated Streams.* Plenum Press. New York. 398 pp.

Way, D. 2003. Beating the Heat: Mitigating Thermal Impacts. *Nonpoint Source News-Notes.* 72:23-26.

Weatherbe, D.G. 1995. A Simplified Stream Temperature Model for Evaluating Urban Drainage Inputs. In *Modern Methods for Modeling the Management of Stormwater Impacts*. James, W. Ed. Computational Hydraulics International, Guelph, Canada. pp. 259-273.

Wehrly, K.E.; Wiley, M.J.; Seelbach, P.W. 2003. Classifying Regional Variation in Thermal Regime Based on Stream Fish Community Patterns. *Transactions of the American Fisheries Society.* 132:18-38.

Wehrly, K.E.; Wiley, M.J.; Seelbach, P.W. 1998. *Landscape-Based Models that Predict July Thermal Characteristics in Lower Michigan Rivers.* Report 2037. Michigan Department of Natural Resources, Ann Arbor, MI.

Wehrly, K.E., 1999. *A Thermal Habitat Classification for Lower Michigan Rivers.* Report 2038. Michigan Department of Natural Resources, Ann Arbor, MI.

Winer, R. 2000. *National Pollutant Removal Performance Database for Stormwater Treatment Practices, 2nd Edition.* Center for Watershed Protection, Ellicott City, MD.

Xie, D.M. 1993. Modeling Solar Thermal Enrichment of Streams Due to Urban Storm Water. MS Thesis. University of Guelph, Guelph, Canada.

WASTEWATER UTILITY

Alabama
Montgomery Water Works & Sanitary Sewer Board

Alaska
Anchorage Water & Wastewater Utility

Arizona
Gila Resources

Glendale, City of, Utilities Department

Mesa, City of

Peoria, City of

Pima County Wastewater Management

Arkansas
Little Rock Wastewater Utility

California
Calaveras County Water District

Central Contra Costa Sanitary District

Contra Costa Water District

Crestline Sanitation District

Delta Diablo Sanitation District

Dublin San Ramon Services District

East Bay Dischargers Authority

East Bay Municipal Utility District

El Dorado Irrigation District

Fairfield-Suisun Sewer District

Fresno Department of Public Utilities

Irvine Ranch Water District

Las Virgenes Municipal Water District

Livermore, City of

Lodi, City of

Los Angeles, City of

Los Angeles County, Sanitation Districts of

Napa Sanitation District

Orange County Sanitation District

Palo Alto, City of

Riverside, City of

Sacramento Regional County Sanitation District

San Diego Metropolitan Wastewater Department, City of

San Francisco, City & County of

San Jose, City of

Santa Barbara, City of

Santa Rosa, City of

South Bayside System Authority

South Coast Water District

South Orange County Wastewater Authority

Sunnyvale, City of

Union Sanitary District

West Valley Sanitation District

Colorado
Boulder, City of

Colorado Springs Utilities

Littleton/Englewood Water Pollution Control Plant

Metro Wastewater Reclamation District, Denver

Connecticut
The Mattabasset District

New Haven, City of, WPCA

District of Columbia
District of Columbia Water & Sewer Authority

Florida
Broward, County of

Fort Lauderdale, City of

Gainesville Regional Utilities

JEA

Miami-Dade Water & Sewer Authority

Orange County Utilities Department

Orlando, City of

Reedy Creek Improvement District

Seminole County Environmental Services

St. Petersburg, City of

Stuart Public Utilities

Tallahassee, City of

Tampa, City of

Toho Water Authority

West Palm Beach, City of

Georgia
Atlanta Department of Watershed Management

Augusta, City of

Clayton County Water Authority

Cobb County Water System

Columbus Water Works

Fulton County

Gwinnett County Department of Public Utilities

Savannah, City of

Hawaii
Honolulu, City and County of

Illinois
American Bottoms Wastewater Treatment Plant

Greater Peoria Sanitary District

Kankakee River Metropolitan Agency

Metropolitan Water Reclamation District of Greater Chicago

North Shore Sanitary District

Wheaton Sanitary District

Iowa
Ames, City of

Cedar Rapids Wastewater Facility

Des Moines Metro Wastewater Reclamation Authority

Iowa City, City of

Kansas
Johnson County Unified Wastewater Districts

Unified Government of Wyandotte County/Kansas City, City of

Kentucky
Louisville & Jefferson County Metropolitan Sewer District

Louisiana
Sewerage & Water Board of New Orleans

Maine
Bangor, City of

Portland Water District

Maryland
Anne Arundel County Bureau of Utility Operations

Howard County Department of Public Works

Washington Suburban Sanitary Commission

Massachusetts
Boston Water & Sewer Commission

Upper Blackstone Water Pollution Abatement District

Michigan
Ann Arbor, City of

Detroit, City of

Holland Board of Public Works

Lansing, City of

Owosso Mid-Shiawassee County WWTP

Saginaw, City of

Wayne County Department of Environment

Wyoming, City of

Minnesota
Rochester, City of

Western Lake Superior Sanitary District

Missouri
Independence, City of

Kansas City Missouri Water Services Department

Little Blue Valley Sewer District

Nebraska
Lincoln Wastewater System

Nevada
Henderson, City of

New Jersey
Bergen County Utilities Authority

Passaic Valley Sewerage Commissioners

New York
New York City Department of Environmental Protection

Rockland County Solid Waste Management Authority/ Sewer District

North Carolina
Charlotte/Mecklenburg Utilities

Durham, City of

Metropolitan Sewerage District of Buncombe County

Orange Water & Sewer Authority

Ohio
Akron, City of

Butler County Department of Environmental Services

Columbus, City of

Metropolitan Sewer District of Greater Cincinnati

Northeast Ohio Regional Sewer District

Summit, County of

Oklahoma
Tulsa, City of

Oregon
Clean Water Services

Eugene/Springfield Water Pollution Control

Water Environment Services

Pennsylvania
Philadelphia, City of

University Area Joint Authority, State College

South Carolina
Charleston Commissioners of Public Works

Mount Pleasant Waterworks & Sewer Commission

Spartanburg Sanitary Sewer District

Tennessee
Cleveland, City of

Knoxville Utilities Board

Murfreesboro Water & Sewer Department

Nashville Metro Water Services

Texas
Austin, City of

Dallas Water Utilities

Denton, City of

El Paso Water Utilities

Fort Worth, City of

Gulf Coast Waste Disposal Authority

Houston, City of

San Antonio Water System

Trinity River Authority

WERF SUBSCRIBERS

Utah
Salt Lake City Corporation

Virginia
Alexandria Sanitation Authority
Arlington, County of
Fairfax County Virginia
Hampton Roads Sanitation District
Henrico, County of
Hopewell Regional Wastewater Treatment Facility
Loudoun County Sanitation Authority
Lynchburg Regional WWTP
Prince William County Service Authority
Richmond, City of
Rivanna Water & Sewer Authority

Washington
Edmonds, City of
Everett, City of
King County Department of Natural Resources
Seattle Public Utilities
Sunnyside, Port of
Yakima, City of

Wisconsin
Green Bay Metro Sewerage District
Madison Metropolitan Sewerage District
Milwaukee Metropolitan Sewerage District
Racine, City of
Sheboygan Regional Wastewater Treatment
Wausau Water Works

Australia
South Australian Water Corp.
Sydney Water Corp.
Water Corp. of Western Australia

Canada
Greater Vancouver Regional District
Toronto, City of, Ontario
Winnipeg, City of, Manitoba

Mexico
Servicios de Agua y Drenaje de Monterrey, I.P.D.

New Zealand
Watercare Services Limited

Singapore
Singapore Public Utilities Board

United Kingdom
Yorkshire Water Services Limited

STORMWATER UTILITY

California
Los Angeles, City of, Department of Public Works
Monterey, City of
San Francisco, City & County of
Santa Rosa, City of
Sunnyvale, City of

Colorado
Boulder, City of

Georgia
Griffin, City of

Iowa
Cedar Rapids Wastewater Facility
Des Moines Metro Wastewater Reclamation Authority

Kansas
Overland Park, City of

Kentucky
Louisville & Jefferson County Metropolitan Sewer District

Maine
Portland Water District

Minnesota
Western Lake Superior Sanitary District

North Carolina
Charlotte, City of, Stormwater Services

Pennsylvania
Philadelphia, City of

Tennessee
Chattanooga Storm Water Management

Washington
Seattle Public Utilities

STATE

Arkansas Department of Environmental Quality
Fresno Metropolitan Flood Control District, Calif.
Urban Drainage & Flood Control District, Colo.

CORPORATE

ADS Environmental Services
The ADVENT Group Inc.
Alan Plummer & Associates
Alden Research Laboratory
Aqua-Aerobic Systems, Inc.
Aquateam–Norwegian Water Technology Centre A/S
BaySaver Inc.
BioVir Laboratories, Inc.
Black & Veatch
Boyle Engineering Corporation
Brown & Caldwell
Burns & McDonnell
CABE Associates Inc.
The Cadmus Group
Camp Dresser & McKee Inc.
Carollo Engineers Inc.
Carpenter Environmental Associates Inc.
CDS Technologies Inc.
Chemtrac Systems Inc.
CH2M HILL
Damon S. Williams Associates, LLC
David L. Sheridan, P.C.
Dewling Associates, Inc.
Earth Tech Inc.
Eco-Matrix
Ecovation
Environmental Engineers International
EMA Inc.
The Eshelman Company Inc.
Finkbeiner, Pettis, & Strout Inc.
Freese & Nichols, Inc.
ftn Associates Inc.
Fuss & O'Neill, Inc.
Gannett Fleming Inc.
Golder Associates Inc.
Greeley and Hansen LLC
Hazen & Sawyer, P.C.
HDR Engineering Inc.
HNTB Corporation
HydroQual Inc.
Infilco Degremont Inc.
Ingersoll-Rand Energy Systems
Insituform Technologies Inc.
Institute for Environmental Technology & Industry, Korea
Jacobson Helgoth Consultants Inc.
Jason Consultants Inc.
Jordan, Jones, & Goulding Inc.
KCI Technologies Inc.
Kelly & Weaver, P.C.
Kennedy/Jenks Consultants
Komline Sanderson Engineering Corporation
Lawler, Matusky & Skelly Engineers, LLP
Limno-Tech Inc.
Lombardo Associates Inc.
Malcolm Pirnie Inc.
McKim & Creed
MEC Analytical Systems Inc.
Metcalf & Eddy Inc.
MWH
New England Organics
Odor & Corrosion Technology Consultants Inc. (OCTC)
Oswald Green, LLC
PA Government Services Inc.
Parametrix Inc.
Parsons
Post, Buckley, Schuh & Jernigan
R&D Engineering/Conestoga Rover & Associates
The RETEC Group
RMC, Inc. (Raines, Melton & Carella)
R.M. Towill Corporation
Ross & Associates Ltd.
Royce Technologies
SAIC Maritime Technical Group
Short Elliott Hendrickson, Inc.
Stantec Consulting Group, Inc.
Stormwater Management, Inc.
Synagro Technologies, Inc.
Tetra Tech Inc.
Trojan Technologies Inc.
URS Corporation
Veolia Water NATC
Wade-Trim Inc.
Weston Solutions Inc.
Woodard & Curran
WRc/D&B, LLC
WWETCO, LLC
Zoeller Pump Company

INDUSTRY

American Electric Power
ChevronTexaco Energy Research & Technology Company
The Coca-Cola Company
Dow Chemical Company
DuPont Company
Eastman Kodak Company
Eli Lilly & Company
Merck & Company Inc.
ONDEO Services
Procter & Gamble Company
PSEG Services Corp.
RWE Thames Water Plc
Severn Trent Services Inc.
United Water Services LLC

Note: List as of 7/1/04

WERF Product Order Form

As a benefit of joining the Water Environment Research Foundation, subscribers are entitled to receive one complimentary copy of all final reports and other products.Additional copies are available at cost (usually $10).To order your complimentary copy of a report, please write "free" in the unit price column.WERF keeps track of all orders.If the charge differs from what is shown here, we will call to confirm the total before processing.

Name ______ Title ______

Organization ______

Address ______

City ______ State ______ Zip Code ______ Country ______

Phone ______ Fax ______ Email ______

Stock #	Product	Quantity	Unit Price	Total

Postage & Handling	
VA Residents Add 4.5% Sales Tax	
Canadian Residents Add 7% GST	
TOTAL	

Method of Payment:

(All orders must be prepaid.)

❑ Check or Money Order Enclosed

❑ Visa ❑ Mastercard ❑ American Express

Account No. ______ Exp. Date ______

Signature ______

Shipping & Handling:

Amount of Order	United States	Canada & Mexico	All Others
Up to but not more than:	Add:	Add:	Add:
$20.00	$5.00*	$8.00	50% of amount
30.00	5.50	8.00	40% of amount
40.00	6.00	8.00	
50.00	6.50	14.00	
60.00	7.00	14.00	
80.00	8.00	14.00	
100.00	10.00	21.00	
150.00	12.50	28.00	
200.00	15.00	35.00	
More than $200.00	Add 20% of order	Add 20% of order	

*minimum amount for all orders

Note: Please make checks payable to the Water Environment Research Foundation.

To Order (Subscribers Only):

Log on to www.werf.org and click on the "Product Catalog."

Phone: (703) 684-2470
Fax: (703) 299-0742.

WERF
Attn: Subscriber Services
635 Slaters Lane
Alexandria, VA 22314-1177

To Order (Non-Subscribers):

Non-subscribers may be able to order WERF publications either through WEF (www.wef.org) or IWAP (www.iwapublishing.com).Visit WERF's website at www.werf.org for details.